辽宁省博物馆学术文库

辽代砖塔雕塑艺术

THE SCULPTURE ART OF BRICK PAGODAS IN THE LIAO DYNASTY

包恩梨 徐秉琨 郑波 著

文物出版社

图书在版编目（ＣＩＰ）数据

辽代砖塔雕塑艺术 ：汉文、英文 / 包恩梨，徐秉琨，郑波著. -- 北京 ：文物出版社，2025.5
ISBN 978-7-5010-6698-8

Ⅰ．①辽… Ⅱ．①包… ②徐… ③郑… Ⅲ．①砖结构－古塔－建筑艺术－研究－中国－辽代－汉、英 Ⅳ.①TU-092.461

中国版本图书馆CIP数据核字(2020)第087133号

辽代砖塔雕塑艺术
LIAODAI ZHUANTA DIAOSU YISHU

著　　者：包恩梨　徐秉琨　郑　波
摄　　影：张守国　郑　波
封面设计：郑　波
版式设计：郑　波　周嘉欣　余　锦
翻　　译：戴洪文　邱楚云　傅启恒　陈柳庄
　　　　　郑　雯　徐　跃　公　睿

责任编辑：乔汉英
责任印制：邵　彬
责任校对：耿瑷洁

出版发行：文物出版社
社　　址：北京市东城区东直门内北小街2号楼
网　　址：http://www.wenwu.com
邮　　箱：wenwu1957@126.com
印　　制：雅昌文化（集团）有限公司
经　　销：新华书店
开　　本：889mm×1194mm　1/12
印　　张：32
插　　页：3
版　　次：2025年5月第1版
印　　次：2025年5月第1次印刷
书　　号：ISBN 978-7-5010-6698-8
定　　价：590.00元

作者简介

　　包恩梨（1930~2008年），山东海阳人，生于吉林梨树县。辽宁省博物馆研究员、中国美术家协会会员、中国考古学会会员，曾任辽宁省知识界女友联谊会副会长。自幼热爱艺术与文化典籍。1953年毕业于中央美术学院绘画系。1956年参加由中央文化部和中国科学院联合组织的黄河水库考古工作队，在三门峡、刘家峡库区做考古调查并参加庙底沟仰韶文化遗址及虢国墓地的发掘。曾主持叶茂台辽墓出土丝织物及服饰品的室内发掘和整理工作、馆藏丝绣文物陈列的设计工作。又负责馆图书室，按照国家部署要求做馆藏古籍版本的调查并对有关古籍进行研究。著有古代绘画研究、文物研究、民俗研究、古籍版本研究等论文多篇。2025年是她九十五岁的生辰，谨出版此书作为纪念。

　　徐秉琨，江苏徐州人，1933年生，辽宁省博物馆研究员，曾任辽宁大学教授（兼）。编著有《北燕冯素弗墓》《东北文化》等专著。

　　郑波，辽宁沈阳人，1957年生，1983年鲁迅美术学院油画系毕业。鲁迅美术学院教授，中国美术家协会会员、油画协会会员。1987年骑自行车考察黄河流域文明。1993年受邀出访美国，为芝加哥奥黑尔国际机场创作大型油画壁画《芝加哥印象》并被收藏。其油画、壁画作品先后在全国美展获奖，多次参加国际艺术展，部分作品被收藏。

Bao Enli (1930–2008), a native of Haiyang in Shandong, was born in Lishu County, Jilin. She was a professor at the Liaoning Provincial Museum, and both a member of the Chinese Artists Association and Chinese Archaeological Society. She once served as the voted Deputy Chair of the Liaoning Female Intellectuals Association. From a small age, she loved art and books on culture. She graduated with a major in painting from the Central Academy of Fine Art in 1953. In 1956 she joined the Huanghe Reservoir Archaeology Team, which was jointly organised by the Central Ministry of Culture and the Chinese Academy of Sciences. Archaeological work was undertaken at the reservoir areas at Sanmenxia and Liujiaxia together with excavation at the Miaodigou Yangshao culture site and Guoguo Tombs. She co-ordinated the excavation of silk articles and clothing from the Yemaotai Liao tomb, and the subsequent work of classifying, storing and designing the displays. In addition, she was responsible for the museum library and collating all the old books according to guidelines set by the government. She has written on a variety of subjects including research on ancient paintings, cultural relics, folk custom and old texts. The year (2025) is her 95th birthday, and this book is hereby published in memory of her.

Xu Bingkun, a native of Xuzhou in Jiangsu, was born in 1933. As a professor at the Liaoning Provincial Museum, the former professor of Liaoning University, he has compiled many monographs, such as *Feng Sufu Couple's Tombs of the Northern Yan*, *The Culture of Northeast China*, etc..

Zheng Bo, a native of Shenyang in Liaoning, was born in 1957. He graduated with a major in oil painting from the Lu Xun Academy of Fine Arts in 1983 and became a school's professor. He is both a member of the Chinese Artists Association and Chinese Oil Painters Association. In 1987, he investigated the Yellow River civilization by bike. In 1993, he was invited to the United States by the Chicago government. In addition, he painted *The Chicago Impression*, a large oil painting for Chicago's O'Hare International Airport. His paintings and murals have won many awards at the national art fair, and have participated in many international art exhibitions. Some of his works have been collected.

编辑说明

以历史学家的视角和责任

这是一部被遗忘的反映一段草原民族文明发展历史辉煌物证的书，更是古代草原丝绸之路文明发展的珍贵历史见证。

徐秉琨先生为了给包恩梨先生生前一部《辽代雕塑艺术》书稿配图，邀请我按照包先生书中需要的图片去寻找、拍照。经徐先生的指导，我们共同努力，克服困难，完成了一个个人无法完成的伟大工程。我们用两年的时间穿越在大辽土地上搜集、挖掘、整理、精选雕塑，归纳、编辑后发现信息量太大。辽代砖塔雕塑又独特，徐先生建议先出以"辽代砖塔雕塑"为主题的一部，这样就可以更早地让大家感受到辽代砖塔雕塑的魅力。我们将继续整理其余书稿以完成包先生全书的出版工作。"辽代砖塔雕塑"是包先生生前研究辽代雕塑的重点，本次又补充了新发现的内容，这更能体现包先生生前研究辽代雕塑完整深研的精神。作为后辈能参与到书中工作，我从心底里感到荣幸。

此书出版，是对中国美术发展史的一大贡献。它将揭开一段不被人注意的辉煌文明。此时，我们的劳动成果即将呈现……我们把收获的这段草原民族与其他民族相融合的辽代砖塔雕塑史奉献给读者。

郑波

2017年12月于沈阳

From the Perspective and Responsibility of Historians

This is a book which reflects the splendid civilization developed by nationalities living in the prairie, and the precious historical proof indicating the civilization developed along the ancient Grassland Silk Road.

Prof. Xu Bingkun was adding pictures for *Sculpture Art of the Liao Dynasty*, a book written by the late Prof. Bao Enli. Invited by Prof. Xu, I took part in the work to find and take photos for this book. Guided by Prof. Xu, we worked together to overcome various difficulties and complete a great project which would not be done on one's own. We spent two years searching, exploring, collecting and selecting sculptures, finding that it contains numerous information after collection and compilation. And sculptures on brick pagodas created in the Liao Dynasty are unique. Therefore, Prof. Xu suggested to publish a book about sculptures of the brick pagodas created in the Liao Dynasty first, then people will enjoy their charming as early as possible. Of course, we would not stop our work until Prof. Bao's book is completely published. This further indicates Prof. Bao's endeavor to study sculptures created in the Liao Dynasty. As the younger generation, I feel honored to participate in such work.

The publication of this book is a great contribution to the development of Chinese art. It will uncover an unnoticed, glorious civilization. Our works will be coming at this time. We will dedicate the history of grassland ethnic and multi-ethnic integration to the readers.

Zheng Bo

December, 2017 in Shenyang

目　录

CONTENTS

序

时光荏苒，岁月如梭，不知不觉间，我十分尊敬的前辈包恩梨先生仙逝近十年了，我也由先生常呼的"小马"变成已过知天命之年的名副其实的"老马"。近些年，不时忆起与先生交往的点点滴滴，感佩先生刚直不阿、追求真理的高贵品格和严谨求实的治学态度，感念先生对我的无私帮助和殷切期望，总想为先生写点什么，但一直未能如愿。前一段时间，先生的爱人，也是我读研究生时的导师，我馆原馆长徐秉琨先生从深圳居所打来电话，嘱我为先生的遗著《辽代雕塑艺术》（《辽代砖塔雕塑艺术》为此书的一章）写序，虽自知才疏学浅，且学业荒废，但仍乐于从命。

先生出生在书香门第，自幼喜爱中华传统文化。1953年毕业于中央美术学院绘画系，曾得到过徐悲鸿等艺术大师的教诲。1956年参加由中央文化部和中国科学院联合组织的黄河水库考古工作队，并与徐秉琨先生结识，因而之后来到辽宁省博物馆的前身——东北博物馆工作，主要从事艺术考古研究。

我在1986年大学毕业到辽宁省博物馆工作后就认识了先生，但开始接触机会很少。1987年9月辽宁省博物馆在甘肃省博物馆举办"齐白石绘画作品展"，我有幸与先生和另两位老师组成代表团同赴西北，其间接触交流很多，从此成了无话不谈的忘年交。我心目中的先生，是一位典型的知识女性，博学强识，秀外慧中，而又豪爽豁达，坦荡正直。治学上善于观察，勤于思考，看问题角度新颖，往往出人意料。发现问题会穷追不舍，刨根问底，韧劲超强。她乐于与他人探讨学术问题，分享学术成果，又勇于坚持己见，为此费尽心力地搜集能支持其论点的可以征信的论据，直到正本清源为止，这也是先生普遍给人留下的深刻印象。先生的牺牲精神同样突出。在馆内，她将大部分精力用于整理编辑图书文献资料，在家里又揽下所有家务以全力保障徐秉琨先生为馆务操劳，个人的治学时间少之又少，且呈碎片化，这也是先生留下的学术成果偏少的原因。但先生的论文都是有质量保证的，其中最大的特点，就是问题导向，聚焦主题，不作空泛宏阔之论，不发无病呻吟之语，以探究学问、答疑解惑为宗旨。如1985年发表于《辽金契丹女真史研究》的《"佛妆"小考》，针对契丹族妇女冬季护面用的"佛妆"进行考证，题目很小，文字很短，但发前人所未发，有学术价值；1983年发表在《社会科学战线》的《〈通志〉版刻考》，对宋元明三朝刊印"三通"之一的南宋郑樵《通志》和《二十略》的情况逐一考证，条分缕析，论述精当；1986年发表于《辽海文物学刊》的《明代画家王守谦的〈千雁图〉及有关问题》，对辽宁省博物馆藏不为人关注的自署"四明锦衣后人"的明代画院画家王谔后人王守谦的生平、艺术和历代以大雁为题材的绘画进行梳理考证，有补画史空白之功；1989年发表在《辽海文物学刊》的《〈山溪水磨图〉试析》，对辽宁省博物馆藏的原定为元代作品的《山溪水磨图》的创作年代，通过画中建筑、道具、人物服饰等，并与传世同类作品相参照，进行综合比较分析，最后得出此图为南宋时期作品的结论。

据我了解，先生长期倾力的学术课题是辽代的雕塑艺术，这一课题选择符合其美术科班出身和一贯的治学理念，她要向又一个学术空白领域进发。

　　源于鲜卑的契丹是中国历史上在文化艺术方面颇有成就和影响的北方骑马民族，他们成功地吸纳了以往的东胡、匈奴、乌桓、鲜卑的文化精华，又积极主动学习借鉴先进的汉文化，取精用宏，消化吸收，改良出新，创造出具有鲜明特色的民族文化，在中华民族传统文化中占有重要地位。契丹艺术丰富多样，其中的雕塑艺术具有突出的本民族文化特征。作为北方草原上的游牧民族，其生存环境、文化传统、生活习俗等与中原明显不同，北方特有的地理环境、生存空间和契丹人的生存方式，塑造了其独特的空间意识和审美取向，体现在雕塑素材、雕饰技法、艺术造型等方面具有浓郁的民族风情，蕴含着丰富的文化信息和艺术养分。在先生所处的年代，限于多种因素，大量的辽代雕塑艺术品尚未被广泛关注，也无人做系统研究，当时出版的多部"中国美术史"或"中国艺术史"专著，在此处均付诸阙如。先生有志于此，自有其专业优势，但面临重重困难。一是没有前人的系统成果可资借鉴，几乎是另起炉灶；二是文献资料很少且语焉不详；三是实物资料分散，许多需要进行田野调查和现场踏勘，在20世纪80年代，难度可想而知。但先生的韧性和爆发力是惊人的，好像她生来就具有应对压力和挑战的特殊本领。她义无反顾地投入这一课题。查阅文献，搜集资料，撰写纲目……先生在默默地发力。1990年退休后更是全力以赴，每每拖着病弱之躯埋首案头。直到2008年去世前在医院病床上，先生还在与我谈这部书稿的撰写。

　　《辽代雕塑艺术》一书分概况、石刻艺术、泥陶塑艺术、砖雕艺术、金属木玉类雕刻与錾琢艺术共五章，以辽代传世和出土的雕塑遗存为实物资料，借鉴美术考古学的研究方法，对辽代石刻、砖刻、木雕、泥塑的题材内容、艺术风格、历史背景进行了系统梳理和全面分析，具有很高的学术价值。《辽代砖塔雕塑艺术》作为全书中重要的一章，此次先行结集出版，是对先生的告慰，对中国美术史的研究是有益的补充，为当代深入研究辽代的雕塑艺术奠定了坚实的基础。这部用心血凝成的书稿，是先生作为一代学人对当代学术的倾情奉献，也是对当代学人的治学示范。作为后学，对书稿感受最深的是字里行间透出的先生浓浓的学术情怀和治学精神，这里有"踏石留印""抓铁有痕"的执着精神，有生命不息、奋斗不止的进取精神，有燃烧自己、照亮他人的奉献精神，有与时俱进、不懈探索的创新精神，这些都是国家正在大力倡导的又是当代人所普遍缺乏的精神。这也是先生这部遗著的特殊价值所在。

　　期待本书的早日面世。

马宝杰

辽宁省博物馆馆长

2017年10月于沈阳

PREFACE

How time flies! It has been close to a decade since Prof. Bao Enli, who I respected deeply, passed away without realising about the time. She used to call me 'Xiao (young) Ma'. Now I become 'Lao (old) Ma' since I have already passed the year of knowing my sacred mission. In recent years, I often recalled the memories of working with Prof. Bao. I admired her noble character of being upright and pursuing truth, as well as her rigorous and pragmatic approaches to research. I felt deeply grateful for her selfless help and ardent expectation on me. I always thought about writing something for her, but the opportunity never arose. Recently, Prof. Bao's husband, my master's advisor called me from his home in Shenzhen. He asked if I could write a preface for her posthumous work *Sculpture Art of the Liao Dynasty*. Although I am undeserving, I am happy to accept the honour.

Prof. Bao was born into a scholarly family. She loved Chinese traditional culture from a young age. She studied painting and received teachings of Great master, Xu Beihong and others at the Central Academy of Fine Art and graduated in 1953. In 1956, she joined the Huanghe Reservoir Archaeology Team, jointly organised by the Central Ministry of Culture and the Chinese Academy of Sciences. She met her husband Mr. Xu Bingkun there. Then they came to work at the Northeastern Museum, the predecessor of the Liaoning Provincial Museum. She continued her research on the archaeology of art there.

I knew Prof. Bao since I came to work at the Liaoning Provincial Museum after I graduated from the university in 1986. We had limited contact at the beginning. However, in September 1987, I, Prof. Bao and two other teachers represented our museum and went to open 'Qi Baishi's Art Exhibition' at the Gansu Provincial Museum, in the northwest of China. In that project, I exchanged ideas with Prof. Bao very often and we became friends who could almost talk about anything without any barrier of age differences. In my mind, Prof. Bao is a typical knowledgable woman with a wide range of knowledge, beautiful and intelligent at the same time. And also she is generous and open-minded, considerate and warm-hearted. As a scholar, she is observant and thoughtful that she always viewed problems from unexpected perspectives. She is generous to share her findings and opinions. When she found problems, she would endeavour to seek the truth with strong perseverance. She is well-known to be a scholar who loves discussing research questions, generously sharing her research results backed with strong evidence. Her fine spirit of sacrifice and devotion to duty will ever be remembered. At the museum, she focused on managing, organising and conducting researches on the collection of books, including ancient rare books with the best qualities. She managed the household so that her husband could concentrate at work in the museum. This caused her only have limited fragmented time for doing her research and didn't publish her scholarly works much. But, her published dissertations and essays are all guaranteed to be of high quality and in-depth studies on specific topics. For example, she published 'Archaeological Research of Fo Zhuang (Buddhist Makeup)' on *Historical Studies on Khitan Women in Liao and Jin Dynasties* in 1985, where she shared her findings of a type of Khitan female face protection cream to protect skin from the cold temperature in the wintertime. The cream has a golden colour, like the skin colour of the Buddha. Before her, there are no other scholars have researched in this unique topic. In 1983, she published 'A Study on Version Engraving of Tong Zhi' on *Social Science Front*. In the article, she shared her thorough research and discussions on the 'Tong Zhi (chorography)' written by Zheng Qiao in the Southern Song Dynasty and 'twenty summaries'. In 1986, she published 'The Ming Wang Shouqian's Painting of Thousands of Wild Geese and Related Issues' on *Liaohai Journal of Chinese Cultural Relics*. It includes her research of the artist's life, artworks and the artworks of wild geese in the successive dynasties. The painting is collected by the Liaoning Provincial Museum. Her findings filled a major historical research gap. In 1989, she published 'An Experimental Analysis of the Painting Mountain Water Mill' on *Liaohai Journal of Chinese Cultural Relics*. In that article, she made a strong argument about the time of the making of this masterpiece. She argued that the work was created in the Song Dynasty instead of Yuan, because of the architectures, props, apparels and so on. She also compared it with other relevant paintings to support her argument.

From what I know, Prof. Bao has focused on researching sculptures of the Liao Dynasty for a long time. This subject is in line with her profession of art studies as well as her pioneering spirit of exploring an undiscovered research area again.

The Khitans, originated from Xianbei Clan, are famous for their culture and art as the northern tribe on horses. They absorbed the best from the past Eastern Hu, Hun, Wuhuan and Xianbei Cultures. They are also actively embracing and learning from more advanced Han Culture. They created their folk culture on top of what they've learnt and absorbed with their original flavour. They hold an important place in Chinese traditional culture. Out of the rich and varied formats of art by Khitans, their carving demonstrates prominent features of the local culture. As a tribe of nomads on the northern plain, Khitan's living environment, culture and customs are different from the Central Plains. These shaped their unique sense of space and aesthetics. In Khitans carving works, their choice of materials, carving techniques and styles reflect their folk customs, embodied with rich cultural information and art resources. Back to the days when Prof. Bao was doing such research, she was restricted by the available research materials, since no one before her has done any systematic research about sculptures in the Liao Dynasty. Despite her professional background giving her advantage in pursuing this interest, she encountered countless obstacles. Firstly, no previous systematic research was available, she had to start from scratch. Secondly, there are just a few previous literature reviews and nothing in detail. Thirdly, the surviving artefacts and information were widely scattered, requiring field research and investigation on the sites. As you could imagine how difficult it would be back to the days in the 1980s, this posed serious logistics challenges. Fortunately, Prof. Bao's tenacious spirit and dynamic character are legendary. It is as if she was born with a superpower to confront challenge and pressure. Without hesitation, she proceeded with the research, from gathering information, reviewing literature to drafting an outline. She worked silently through her retirement. Despite poor health, she devoted herself to the endeavour with more fervour after retirement in 1990. She was discussing the manuscript with me even when she was in the hospital bed before she passed away in 2008.

This book, *Sculpture Art of the Liao Dynasty* contains five chapters: Abstract, Stone Carving Art, Clay sculpture, Carving on bricks, Metal engraving, Woodcut and Jade carving. The artefacts selected in this book include both passed down and unearthed cultural survivals from the Liao Dynasty. This book is with immense academic value. Prof. Bao applied art archaeology research methods to conduct a systematic review and holistic analysis on the theme, art style and historic background. The information is gathered from Liao artefacts passed down or unearthed and she applied art archaeology research methods on analysing the findings. This systematic and comprehensive analysis on the subject of Liao stone, brick, wood carving and clay pottery examining different artistic styles and historical background, is of immense academic value. *The Sculpture Art of Brick Pagodas in the Liao Dynasty*, as an important chapter in the book, has been collected and published first, as a consolation to Prof. Bao and a supplement to researches on Chinese art history. It sets up a solid foundation for further studies of carving art in the Liao Dynasty. This book contains hard works from Prof. Bao, represents the academic contribution of her generation. It demonstrates her approaches to studies. For me, as someone from a younger generation, I can feel deeply about her passion for doing studies and rigorous attitudes toward research from reading between the lines. This embodied her tenacious spirit and determination to proceeding evidence-based research regardless of difficulties and challenges. She is like a candle that she sacrificed herself to ignite others. She progressed with the times fearlessly, explored and innovated relentlessly and made breakthroughs again and again. These are exactly what the nation is actively promoting but sadly it is also what the majority of people in our current generation are still lacking for. Her spirit is also where the distinctive value of this book lies.

I eagerly await the publication of this manuscript.

Ma Baojie
Director of Liaoning Provincial Museum
October, 2017 in Shenyang

辽代砖塔的雕塑艺术

辽代是中国历史上继唐代之后崛起的一个朝代。辽王朝是以契丹族为主建立的一个北方民族国家。从公元907年建国至1125年灭亡，两百余年间大致与中原的五代和北宋王朝相终始。辽王朝崇尚佛教，在北方建造了众多的寺庙与佛塔，其砖构佛塔上精美的雕塑装饰为中华民族留下了一笔宝贵的文化财富。

这里，让我们先简略地回顾一下中国砖雕艺术的发展情况。

中国砖雕艺术的诞生年代大约可以追溯到战国时期，也就是说，砖雕艺术几乎是与砖的实用价值同时产生。相关的建筑材料瓦和瓦当也是在那个时候出现的。

早期的砖雕艺术表现技法单调，多是在砖坯烧制前用工具直接将纹饰图样刻划，或以印模压印在坯面上，再经过装窑烧制成为画像砖。这种做法大致同于瓦当而略为复杂。虽然画像砖的内容很丰富，但是表现技法则止于绘画与雕刻之间的在坯面上的线雕与平刻。浅浮雕、高浮雕的表现技法出现于砖雕已是东汉中、晚期。

到了辽代，砖雕艺术之葩开放得更为灿烂。在东北地区契丹民族的发祥地，从考古发现可知，早在秦汉时期就已经有了画像砖艺术的存在。辽代砖雕艺术开始也是承循传统画像砖那种绘画与雕刻之间的阴文线雕与浅浮雕技法，即宋人《营造法式》的雕镌制度中称作"素平""压地隐起""减地平钑"等技法，而随后又出现了"剔地起突"即类同于高浮雕的做法。由这些传统的砖雕技法做成的画像都已出现在辽代砖塔以及墓葬里。不但如此，在辽代砖雕艺术中，还产生了一种最为光彩的、创新的大型组合式的砖雕形式。辽代最丰富的砖雕遗存保存在砖塔建筑中。

塔的建造本来起源于印度，又名"窣堵波"或"塔婆"，为梵语Stupa与巴利文Thupa的音译。开始为供奉佛骨之用，后来也用来供奉佛像或存储僧人遗骨、贮藏经卷。塔柱原是石窟寺造窟时留以支撑窟顶的中心石柱，后来由石窟中分离出来，演变成以土、石建造的独立的建筑形式"塔"。中国的佛塔发展为将原来存放佛物的部分"窣堵波"缩小为塔刹，安置在塔的最高处塔顶之上，而扩展塔身以为登临瞭远之实用。以砖建造的塔是北魏时才有的。出于宗教宣传的需要和装潢美丽的目的，又接受了前代画像砖的技艺传统，利用砖坯本身适于雕琢，可以在烧制前捏塑或模制，又可以在烧成以后再进行磨琢加工，具有比石刻更大的艺术创造自由的特点，便产生了中国特有的佛教艺术形式——砖塔的雕塑艺术。又随着时代的发展，出现了不同时代、不同地区的各自不同的艺术风格。

现存年代最早的佛塔是北魏正光元年（520年）所建河南嵩岳寺密檐十二边形砖塔，已利用砖雕艺术装饰塔身，以后各代砖塔也有砖雕存在。早期的塔雕尚未远离西方的原有塔式，塔身以素面为多，往往仅利用砖雕技法制作起装饰作用的塔檐、门、窗而已。造型拘束，艺术性不强。到了唐代，佛教盛兴，佛塔的建筑艺术与雕塑艺术也都有了创新。其中，河南安阳唐咸通年间（860~874年）建造的修定寺塔便已出现了精美华丽的砖雕。

至辽代，砖塔的建筑与砖塔上面的砖雕艺术达到了创造的高峰。崛起于北方的辽王朝，崇尚佛教，辽太祖耶律阿保机早在称帝之前就已崇佛，唐天复二年（902年）在其任遥辇痕德堇可汗本部夷离堇时，便在龙化州建开教寺（《辽史》卷一《太祖纪上》）。神册三年（918年）阿保机称帝后不久，即诏令修建孔子庙、佛寺、道观（《辽史》卷一《太祖纪上》）。至兴宗耶律宗真（1031~1055年）在位时"尤重浮屠（即塔）法，僧有正拜三公、三师，兼政事令者，凡二十人。贵戚望族化之，多舍男女为僧尼"（《契丹国志》卷八）。辽代的五京，寺院林立，经幢佛塔，遍布境内。至今仍有"唐修寺、辽造塔"的民谚流传。

辽塔的兴建给予砖雕艺术以发展机会。在结构上，辽代砖塔，主要是密檐塔，为仿木建筑，内部为木结构框架，而在外部则以砖砌实，用砖雕制佛、菩萨、弟子的造像，表现佛传故事与供养人的画像及图案纹饰、动物形象。重视偶像崇拜的契丹民族创造性地将佛塔的建造与佛像的供奉有机地结合在一起，使塔身雕塑与塔的建筑浑然一体，不可分割。辽塔的造型，包括它的雕塑艺术在内，也因此在历史上成为后世金、元、明、清以来造塔之定式楷模，不过后世之佛塔建筑已经是它的余韵了。

建于辽天庆十年（1120年）的北京天宁寺塔（插图1）高57.8米，塔身上浮雕"礼佛图"，佛、菩萨形象采取大型绘画式展开画面，用浅浮雕加堆塑技法表现，人物、坐骑逶迤塑来，人在塔下仰望仿佛观赏长卷画轴。

山西灵丘县觉山寺塔须弥座壶门内佛像、壶门之间及塔角力士，形象生动精美。

内蒙古呼和浩特东郊约辽道宗时（1055~1101年）所建万部华严经塔高45.18米，塑有天王、力士、菩萨形象。须弥座及栱眼壁砖雕宝相、牡丹花最为生动（插图2）。

巴林右旗辽庆州遗址内辽重熙十八年（1049年）所建八角七层砖塔（庆州白塔）高49.48米，每一层都嵌有佛、菩萨、力士及乐舞、供养人饮宴场面的画像砖。

插图3
佛像
大明塔

宁城县辽中京遗址内之大明塔高达74米，塔身第一层砖雕造像，佛像面容丰腴，有唐塑遗风（插图3）。辽代帝王学唐比宋，如辽圣宗崇拜唐玄宗李隆基，将自己的名字改为"隆绪"。辽中京是圣宗时所建，大明塔主体佛像力仿唐作，是可以理解的。

辽宁朝阳市云接寺塔、北塔，都是比较早期的辽塔建筑，砖雕艺术风格与造型都留有浓厚的唐代遗风。佛与菩萨以及须弥座壶门内伎乐、供养人的面型、体态都是表现为丰满肥美的"周家样"唐代仕女形象，而北方契丹人物的形象则仅见于壶门外侧供养童子，显有北方儿童状貌。还有天津蓟县白塔、盘山天成寺塔，北京房山云居寺北塔、戒台寺塔、多宝佛塔，辽宁喀左县精严禅寺塔、义县城内嘉福寺塔等等，也都各有特色。

辽代修建的塔是我国古代建筑遗存中最丰富的一部分，辽塔砖雕是继承唐代而又具有契丹自己的时代、地区、民族形式与风格特点的造型艺术。

辽塔大多坐落在两座高山双峰之间的驼鞍部位，下依长河，东迎旭日，以其雄伟挺拔的身姿与浑厚而又瑰丽的艺术形象形成一曲古代民族建筑林苑中的绝唱。在以天地山川为背景陪衬下，其艺术魅力何尝不可以和浩浩江水、皑皑白云同流。曾经千万里广袤辽阔的辽国领域，历经千百年风雨兵火的洗礼，遗存下来矗立着的辽塔，可供我们研究，并从中得到对于辽文化成就的进一步认识。这里试就现存地面的一些具有代表性的辽塔，探讨一下辽代砖雕工艺的特色和它的艺术成就。

一　辽代砖塔的砖雕类型和表现技法

辽代建筑砖塔时镶嵌砖雕的部位和垒砌方法，就目前所见有三种情况。

（1）立砌单砖型。系在烧制前的砖坯表面上，以一砖为一个独立完整的画面，用花模压印出图像和花纹，经过烧制后再磨琢加工，然后画面向外立砌于塔身或须弥座外壁。

（2）连砖式连续花纹型。系以两块长方形砖夹一块正方形砖平列为一组，或四块长方形砖平列为一组，连续组成一个单元画面。或多块砖相连，组成多种连续图案。如莲瓣纹或连珠纹、卷草纹、卷云纹等，常作边缘装饰之用。

（3）组装式人物图案型。系按照图案和造型需要，以坯模制成大、小、方、圆等形状不一、角度不同的小砖砖坯，烧成出窑后再按预先设计的图样加以拼联，组合成完整的人像或花纹图案单元的组合砖雕，成型后镶嵌砌入塔身、须弥座各面以及四周各角。这种类型的组合砖雕大多为高大的佛、菩萨、天王、神兽等高浮雕与半圆雕。这种大型组合砖雕本身是辽代艺术家将建筑艺术与雕塑艺术加以结合的一种创新。辽宁海城市析（sī）木镇辽"铁塔"，塔身六面各以约7厘米厚的卧砖（平置之砖）垒至28～30层，砌成砖雕菩萨，与塔身浑然一体。这种类型在辽塔中以辽宁一带最多。

关于砖雕的技法，与辽代同时的中原汉族地区，佛塔建造的砖雕往往是以阴文线刻与浅浮雕形式表现在以泥灰抹平的大型平面上。而辽塔则运用浮雕、半圆雕、圆雕等技法来突出表达造型的体积感，着重表现被刻画对象的自然风动感、质感、空间感以及它们的活动力度。如辽宁朝阳云接寺塔塔身转角处的负塔力士、海城析木镇铁塔二层西北角残留的失头力士，都生动地证明了这一点，当同一时期宋代的雕塑风格趋向于程式化，表现技法拘于营造制度已呈刻板的时候，辽代砖塔砖雕的活泼艺术风格却显示了特有的醇厚和隽永的魅力。

二　简单介绍几座辽塔砖雕

1.云接寺塔

位于辽宁省朝阳市东南15千米凤凰山云接寺西侧，原名摩云塔。该塔为方形十三级密檐式，全高37米。单层须弥座，束腰每面中间砖雕仿木假门，门上梅花式门钉三排，每排六个。铺首上刻门锁。假门两侧各雕三个壶门，壶门内用砖组装浮雕佛、菩萨、伎乐人，壶门外为化生童子、供养人及花卉、卷草纹、云纹的边饰、角饰等。四角组装半圆雕、圆雕的天王力士各一身。塔身四面的砖雕也是嵌砌组装的高浮雕与浅浮雕的佛、飞天、菩萨、小灵塔。佛与灵塔的上方有宝盖，灵塔上方又有一对小飞天，菩萨头上则无宝盖。

塔身的四面坐佛代表佛教密宗五方如来中的四尊。东面阿閦（chù）佛，五象趺坐。南面宝生佛，五马趺坐。西面无量寿佛，五孔雀趺坐。北面为妙生佛，五金翅鸟趺坐。每面坐佛身侧均侍立二菩萨，戴花冠、璎珞，足踏莲花。大日如来在塔中央，外面不作表示。小灵塔为十三层，塔身浅浮雕佛坛、坐佛。灵塔旁有砖刻榜题文字。东面灵塔左题"庵罗卫林维摩塔"，右题"安婆罗林中圆寂塔"；南面灵塔左题"菩提树下成佛塔"，右题"净饭王宫生处塔"；西面灵塔左题"给孤独苑名称塔"，右题"鹿野苑中法轮塔"；北面灵塔左题"耆暗崛山般若塔"，右题"曲女城边宝阶塔"。檐下斗栱也颇具特点，泥道栱上之慢栱均作鸳鸯交手栱，转角铺作则是缠柱造，同于山西大同华严寺的斗栱做法。

五方如来、八大菩萨、八大灵塔，是辽塔常见佛本生故事主题。在朝阳，这样的辽代砖塔还有几座，一座在凤凰山下的王秃子沟，另外的在朝阳市内。其中市内之"北塔"，原本建于唐代，据考古发现该塔现为辽代包砌外层，塔心内有唐代砖石、佛像等遗存。云接寺塔造型及砖雕都仿佛是北塔的翻版。但在须弥座南面第二壶门外，有屈膝持罐、深目厚唇的典型唐代胡人形象；第五壶门外右侧有戴高帽、着左衽袍服、执骨朵的契丹男供养人形象出现（插图4）。

中原地区自盛唐时期以来以砖塔模仿木结构式样，塔的平面很多是正方形，而且多数朴素无饰。辽金时代佛塔平面大都为八角形，只有少数方形，云接寺塔方形，却又在塔身

插图4
执骨朵供养人
云接寺塔

及须弥座上有众多精美雕饰，应视为是在二承唐风的基础上已经逐渐开始形成辽代独特风格的过渡时期的塔式，是辽代佛塔艺术形成的中早期的代表作品和里程碑。

2.析木镇铁塔

位于辽宁省海城市析木镇路旁，原有铁塔寺早已荡然无存。旧《海城县志》："县城东南四十里析木城西有古塔一座曰铁塔，塔七层，高三丈余……塔下有庙曰铁塔寺。"据析木中学几何教员毛凤举同志于20世纪60年代利用投影测量方法测出塔高为23.6米。1976年海城地震，塔刹塔顶震毁，据久居塔侧的姚荣国同志回忆：地震之前曾于塔顶生一棵高粱，至秋，高粱穗恰好与塔刹齐高，高粱一般高3米左右，故估计铁塔塔刹高3米。当地人称此塔顶为"小庙"塔顶。坠地时有砖雕小佛一身从中坠出，一尺余长短，塔顶坠地粉碎后散砖拉一马车有五六百块。砖与小佛均于地震后散失。现在残塔余高仍约20米。

铁塔为砖筑六角密檐七级塔，塔座于1984年以水泥加固，原貌无存。此塔有两层塔身，第一层塔身下为复瓣仰莲座，形制特殊，为三片莲瓣中间托一簇砖雕莲蕊。莲瓣上即承托高约5米的第一层塔身。塔身六面，每面砖雕组砌高浮雕菩萨一身。菩萨头上出半圆雕弧形五垂绶宝盖，微超出菩萨头轮廓之外，恰好遮罩于菩萨头上。菩萨砖雕以卧砖28层垒砌，每层砖高7厘米，菩萨总高2米左右。菩萨背地以一立砖（29厘米×18厘米）三卧砖（20厘米×7厘米）砌砖壁。每面塔身转角处用一方砖、三卧砖砌成六角形素面立柱（只露半面的角柱）。菩萨形象宽颐丰颔、束发冠髻、宽袍广袖，腰系宽松大带，项下华绳直垂脚面，下有两朵重莲各擎托菩萨一足。

东面菩萨裸右肩，衣披左肩搭左臂侧身向东南方站立，左腕环钏、露左耳珰、双手合十或拱手持物，模糊不清。东南面菩萨合十当胸，身袭袈裟、广袖及膝，衣裾流苏齐垂足下莲朵之上，腰腹下露衣带两条止于广袖下端。无璎珞，背后塔壁左右隐现浅浮雕华绳如横飞状。双耳环珰、冠髻束发，冠饰花纹为辽代雕塑、壁画上常见的卷云头圆花朵，是铁塔六身菩萨中服饰最简而又最宽厚者（插图5）。

西南面菩萨向左侧身而立，弯曲右腕擎持柱状物，左手腕钏下垂、抚持衣带。当胸项圈中衔一花，花下华绳一挂直垂脚面，飘带经左肩搭绕右肘直垂一端至足，露右耳，耳坠环珰，冠髻束发，冠饰三花。西面菩萨头向右倾，露右手，耳坠环珰，冠髻束发，冠饰卷云头间一花，颈饰项圈，左手持双盏，右手抚盏上物，袈裟经右肩绕搭左肩，经左臂垂直于左膝，双足所踏重莲残损，衣裾垂流苏下覆足面。西北面菩萨冠髻束发，冠饰中间圆花一朵，侧首向右，双手捧持重盏两对，袈裟绕体搭左腕垂至膝部，双足之上拂落衣裾流苏，厚一层砖。菩萨身后塔壁上现浅浮雕光环。

东北面菩萨束发冠髻，冠饰中间有圆花一朵，上抬右臂，左手抚腰间衣带，袈裟披右肩，大袍，广袖，腰带自然宽松，于腹前绕系，带的两端飘拂于左右膝前，袍下流苏垂至足面。六身菩萨中，以东北面菩萨面目保存最为完整，又最具宽颐、丰颔的契丹人面型、服饰特点（插图6）。

菩萨宝冠形制与山西大同华严寺泥塑相似，服饰也近于契丹北方御寒常服，与辽庆陵出土木俑及叶茂台女尸所着大袍相似，未加繁缛纹样与更多佩饰，这是西方佛教传入中土以来佛装的中国化和北方民族化的表现。少一些一般菩萨的女性柔美，身躯竟如北方大汉，是此塔菩萨特点。

第一层塔檐之上为一平座，平座上承第二层塔身。塔身甚矮，只等于是一层塔檐与二层塔檐之间的一段壁面。在此壁面上，每面原立砌方砖四块一字横列，砖面各模印浮雕坐

插图5
东南面菩萨
析木镇铁塔

插图6
东北面菩萨
析木镇铁塔

佛一尊，佛身背后有放射形佛光。方砖面约38厘米×38厘米。坐佛像座高占佛高约1/4。每面应四尊佛像，今只残留方砖东面两块、东北面一块、西北面一块、东南面一块。在二层塔身西南角残留负塔力士一身，头部缺损，残高仅30余厘米。着宽袍、帛带束腰，双手互扣握于胸腹前，弓步站立，宽袖扬袂于身后两侧，动作豪放，气魄雄浑，衣饰形制极简单，但是砖雕作者却有意在腰带上浅雕下布帛勒系出现的褶纹以示腰带的织物质地，这是典型的北方游牧民族骑士特有的服饰装束（现在二层塔身各转角处皆有力士，力士亦皆有头，系后来修补，已非原貌）。铁塔无建筑年代见于记载，但是塔砖有6~10道不等的沟纹，为辽砖特点。

铁塔这种二层塔身的建筑形式罕见。按山西应县木塔为五层塔身，第一层塔身为重檐建筑，两层塔檐之间为一段短壁，可悬匾额，故全塔共有六层塔檐。而铁塔虽为密檐，其第一层是否也可视为重檐建筑。然则第二层矮短塔身则仅是重檐建筑的一部分。就塔檐而言，仍是"七级浮屠"。

3.金塔

位于辽宁海城析木镇西北5千米的羊角峪西山，在两峰之间的平岗上，前临析木河。

金塔为八角密檐十三级，全高31.5米，塔的基本方向为正南正北（方向角南偏东2度）。在山石上起砖构台基，叠涩内收约十八层砖，台基周壁每面雕六朵下垂如意云头纹。

塔座为须弥座两层，都有半圆雕复瓣仰莲的平座层，用模压莲瓣各部位形状而成的砖，组合叠砌成片片微微舒展的莲瓣，瓣腹饱满，似乎蕴含着一派生机。须弥座的上下枋都以长框连珠纹饰边，两端饰卷云头。下层须弥座的束腰，每面中间有立砌方砖雕制的瘿项柱，柱身为牡丹图案，柱的上端雕莲瓣，柱脚雕莲叶。花纹突起较低，最高点都与瘿项柱面齐平，图案规则，与宋代石刻"剔地起突"技法相类。瘿项柱两侧有壶门，门内外组合砖雕伎乐人、舞俑、侍者等供养人，八面共计48身。惜人物面部已无一完好，身躯也多有残毁，仅西北、西南、东北、东南及正北五面砖雕保存稍好，但是残留部分仍然能显露原来造型技艺的娴熟，在现存辽代砖塔中应称上品。

供养人形态多样，风格活泼，有结裤管佩璎珞捧物奉献者，坐于硕大过人之酒瓮前者，抱瓮、负瓮者，一派蹒跚醉态，跃然塔壁。尤其以着契丹服装、高靴长袖、回身旋舞的四个男舞人造型最具民族风格与时代特色，几乎与传世辽代绘画之人物服装、舞蹈动作相同而更觉生动，简直是一幅宋代姜夔《契丹风土歌》的画面。束腰西北面壶门两侧砖雕有分别肩负"人""我"两个大字之两名化生童子，字形占据整个画面，肩字童子拘谨于字的笔画中间。东南面砖雕裸臂吹笛与弹琵琶女乐人，虽头部已残损，但是身姿动态仍然优美动人。西南面另有着契丹装之奏琴男乐人，其所抱之琴箱形状与今天仍流行于蒙古族的马头琴形状极为相似，应为马头琴的早期样式。残留的乐人头部仍为动作生动的边弹边唱的说唱艺人姿态。须弥座束腰转角处雕负塔力士，皆鼓腹怒目跪伏作用力背负状，形态生动，亦皆残毁，只有西南角与东北角力士尚清晰，保存较完整，张口侧颈，仿佛力疾负重，极其生动逼真。

上层须弥座束腰每面是一宽大的壶门，门内组装卧砌高近1米的砖雕护法狮子一躯，正面正身，披鬣狞目，顿足生威，胸臂肌肉偾张，探身于须弥座外。四正面（南、北、东、西）各狮前肢双撑，四隅各狮前肢一直支一斜撑。狮身左右宽80~90厘米，高70~77厘米，用10~11层砖拼组砌成。狮眼镶嵌物早被挖窃，却仍威势不减。在辽代石刻李进石棺（现藏辽宁省博物馆）外壁也出现有气势森严的半圆雕形式的"虎""龙"形象，给人的感觉是，金塔砖雕狮子的浑厚艺术风格，与李进石棺之石刻都表现了东北契丹族的写实作风。

上层须弥座的负塔力士仅以后背着附塔身，几乎全身都以圆雕技法表现。凌风凛立，映衬天空而形成的外轮廓十分壮观。但亦只余东北角力士基本完整。

上层须弥座之上施复瓣莲座，承塔身。塔身高3.3米，底边3.35米，各角倚柱柱头卷杀。每面造眉栱龛，龛内雕坐佛一尊，坐八角形须弥座上，半圆砖雕。座甚大，前身已突出龛外。龛外左右立侍菩萨各一身。菩萨以10~20层卧砖装砌，全身半圆雕，以4层卧砖圆雕头部探出塔身，每身菩萨高约1.5米，皆冠带璎珞，足踏莲花。东面菩萨秀面丰颐有唐代风格，而佛与菩萨头多毁。佛龛之上雕砌四垂绶大宝盖，菩萨头上为小宝盖，盖之顶心饰一莲花如器物盖上之纽。

塔檐下，阑额之下，每面佛顶宝盖上有组装卧砌飞天一对飞翔于云端，八面共飞天十六。或立或飞，曲臂持花，欹首合掌。南面飞天作天女散花姿态，体躯面目略失丰肥，身下足端云朵，衣带流逸飘动感很强。砖雕外层敷以很厚的圬泥，是经过后来重修的结果，失去原有风格。西北面飞天，坐云上，作乐舞状。东面飞天乘云朵上昂首俯身翩翩起舞。东北面飞天扬臂侧身飘带翻飞，云朵自半身处遮绕，舞姿身手动作很大。西南面飞天动势相似，但是造型准确，动作自然，头饰具辽代特点，后世重修时未作重大变动。北面飞天在云朵中作遨飞状，侧身卧姿枕臂交足，背向外，卧云朵上，如谛听佛在说法，神态潇洒。这些飞天虽不及敦煌飞天之仙姿纤美，而独具活泼不拘之自如天然。东南面与东北面类似，当据相同的粉本制作。

塔身菩萨及佛像由于后代装修粉圬过厚，大半失掉本来面貌神情，亦失去辽代风格。龛内坐佛更经后世变动，有的已非原物。东面左侧菩萨残损，暴露出背部嵌入塔身之组装卧砖15~16层，右侧菩萨更已脱落无余。只有西北面与北面菩萨由于表面圬粉风化，露出砖地，而充分显露出辽代砖雕的本来面貌。北面菩萨面型、体态在保有辽代风格的基础上，还可以看出有上承唐代重颔丰颐的影响（插图7）。

西北面菩萨以18~20层砖叠砌，与东面左侧菩萨之叠砖层数稍有不同，说明辽代砖雕菩萨虽在一塔之上也可不尽一样，即身高不完全一致。菩萨头冠形式则与大同华严寺辽代泥塑菩萨宝冠形制相同。

塔之北，原有明代建筑金塔寺寺院及明万历三年（1575年）《敕建金塔大宝林寺》碑，万历三十九年（1611年）《重修敕建金塔大宝林禅寺记》碑，万历四十一年（1613年）《重修彩画金塔寺碑记》及清乾隆四十三年（1778年）、五十八年（1793年），嘉庆二十二年（1817年）碑记共六块，今均不存，原金塔寺遗址内现只余龟趺一座。

金塔用砖规格有方砖两种：（1）36.5厘米×36.5厘米×6.5厘米，背有八道沟纹；（2）54厘米×54厘米×7厘米。长方砖三种：（1）23厘米×39厘米×7厘米，背后纵沟两道，中夹三个十字双沟纹；（2）24厘米×41厘米×8厘米，背后抹沟纹九道；（3）18厘米×39厘米×7厘米，背后纵沟纹六道。砖背青色，火候较高，也有少数微红欠火砖，砌入塔内。

4.广济寺塔

位于辽宁省锦州市大广济寺内，据《奉天通志》卷九十二记载，广济寺塔建于辽清宁三年（1057年），曾经明代重修。据考古发现和研究，当年的广济寺原亦为辽代建筑。

塔为砖筑八角十三级密檐式，高57米。塔的基台毁坏，于1983年以小砖加固补修，每面宽8.6米。

塔座现存须弥座一层，束腰高50厘米，每面有砖雕瘿项柱，间隔出5个壶门，壶门内组合砖雕坐佛一尊，壶门外两侧组合砖雕瓶花、供养人。西面砖雕保存完整（插图8），西北面仍清楚可见，其余各面风蚀脱落。须弥座上承仰莲，接两跳五铺作补间斗栱四朵，上托勾栏平座。栏和座已损毁无存。

塔身每面宽8.24米，转角处以砖砌圆形角柱。每面塔身中间一龛，龛内组装卧砖砌圆雕坐佛一尊，坐莲座上，龛外两侧立半圆雕长大于佛身一倍之菩萨。佛龛上方出一横枋，其上罩一具浮雕的宝盖。菩萨头上亦各悬宝盖，大小为佛头宝盖的一半。西南面右侧、东南面右侧、东面右侧菩萨损毁无存。佛与菩萨及宝盖全部卧砖雕砌。塔身正面佛像头戴宝冠，有砖雕盾形背光，其他各面佛像头上螺髻。菩萨宝冠璎珞，头的背后塔壁浅浮雕有圆形佛光，由于灰泥剥落净尽，可以明确看清楚菩萨为小雕砖上下左右连续堆砌成型，组装后整体嵌砌入塔壁上预先留出的位置，安装之后在背地塔壁上浅浮雕出光环。

每面塔身于佛龛宝盖上方有二小飞天，左右斜刺里横身飞着于塔身上角两端。飞天高不过2~3层砖。身姿颀长，与大同华严寺薄伽教藏殿上佛背光边缘上之飞天造型一致。菩萨则高达3米余，砌砖达40层。以小飞天与大菩萨相对，是该塔不同于其他辽塔砖雕之处。

锦州属辽宁滨海区域，空气较湿润，风蚀严重，原来塔壁砖雕残损，原有表面圬粉都已剥蚀。因此，组合砖雕堆砌成型的工艺程序清楚可见，包括雕刻背后附着的木垫与塔身

插图7
北面菩萨
析木镇金塔

木件结构接榫合卯的部分。由于塔身已经由里及表地与雕件联结在一起，如非地震，雕件不易脱落。

广济寺塔菩萨丰颐方颔，身姿茁硕，神态肃穆，与同时的宋代菩萨造型柔美温和不相类，而更接近唐、五代之圆浑、丰满风格。锦州辽置，至今未更名，原为阿保机以汉俘建州，地处东北南部，从地理上看也是中原与东北地区交流的枢纽地带。广济寺塔砖雕造型艺术较析木铁塔具有更多中原风格，是兼有地理与民族原因的。

三　辽塔砖雕的题材特点和艺术特色

辽代密檐砖塔中，平面为方形、六角形者相对较少，大多为八角形。辽塔造型特点是在平面台基上建须弥座，座上置平座（或有仿木砖雕平座斗栱）、仰莲以承托塔身，塔身有一层或两层不等。每面塔身砖雕各式佛、菩萨、飞天等人物形象，以及佛传故事画面，花纹边饰、仿木结构门窗等等。塔身上端为塔檐及檐下斗栱，在多层砖檐之上，塔顶以塔刹作为结束，雕塑艺术与土木建筑紧密结合，成为不可分割的整体。

辽代砖塔不仅形制承袭唐、五代风格，塔身、须弥座上砖雕艺术风格亦保有唐、五代造型艺术传统。朝阳云接寺方塔即以陕西西安香积寺唐代方塔为模范造型，又加以各种雕塑装饰塔身。元、明、清诸朝皆承袭辽塔的八角形形制，而塔身、塔座遍饰雕塑，成为此后中国佛塔的基本样式。这些是早为人们所熟知的了。

然而辽代砖塔的砖雕艺术却是历来少为人道及的。从本书介绍的几座辽代砖塔，可以看到辽代雕塑中砖雕艺术的大概轮廓，同时，可以看到在中国古代雕塑的殿堂里，辽代的砖雕艺术是怎样熠熠生辉的。现就以下几个方面对辽代砖塔之砖雕艺术作一归纳分析。

1.须弥座的发展

须弥座本是一种佛座，但自从出现了塔、幢等佛教建筑，便也以它为建筑底座。辽塔的须弥座是砖雕艺术很大的用武之地，也是辽塔艺术特色所在。

雕塑于塔身的佛、菩萨、飞天等等是神化了的人的形象，但是一经神化就有了它自己的模式。自宋以来，在神像的雕塑方面也有程式的约束及匠人陈陈相因、历代相传使用的"粉本"的约束，虽然在偶像的神情、衣饰上因时、因地小有改动和变化，但究竟是庄严法相，不允许有离题太远的即兴创作。而辽塔在须弥座上的雕塑，就有一定的选题及创作表现的自由。尤其在刻画供养人的题材方面，更加活泼生动。在今天看来还能体味感受到它们的时代气息、民族特点，并且可以从此了解辽代文化面貌的一些方面。

海城析木镇金塔须弥座西南面供养人中长身高歌曼舞的契丹人物的服装动作与传世辽代画家胡瓌所作《卓歇图》中所画舞蹈者的舞蹈动作，以及与许多辽墓壁画中舞蹈人物的服装、动作相似。舞人的双足踏舞、转身回舞更是至今流传于东北地区满、蒙古民族的舞蹈"二人转""地秧歌"中常用的传统舞姿动作。一些舞人长袖交叠、飘舞甩动的动作能够以砖雕表现出如此"动"的效果，表现的角度、转角的关系又交代得极其清楚，也为今人提供了契丹舞蹈的形象资料。唐时的"胡旋舞"是否也如此呢？在河南焦作金元墓葬（见《文物》1979年8期）中出土的"戏乐俑"（插图9），其舞蹈神态、动作、服饰也应属于这种舞式的流传曼衍。

自从塔柱从石窟艺术中解脱独立以来，塔的建筑形式发展到了辽代，在承袭前代佛塔造型艺术的基础上，至此，又开拓了塔的须弥座这块砖雕艺术的新的创作领地。

2.在塔身上突出菩萨的位置是辽塔的特点

辽塔塔身雕塑题材等仍都承袭佛教传统，采用一佛二菩萨的格式，但在布局构图与雕刻手法上有时却更突出菩萨。

海城析木镇铁塔虽然将佛像置于上层塔身，雕塑菩萨于下层塔身，但上层塔身极窄狭，下层塔身宽阔，将菩萨以组装卧砌型做法，作高浮雕雕塑造型，数倍大于佛的单身造像。类似铁塔这样的表现手法，在目前看来尚属孤例。但是从另外几例也可窥见辽代崇尚菩萨的情况。例如塑有一佛二菩萨的塔身，除却朝阳北塔和云接寺塔造像佛大于菩萨，其余往往都是菩萨大于佛。辽宁省铁岭市尚有一座辽代八面密檐砖塔——铁岭白塔，塔身每面砖雕坐佛一尊，而佛头上却为菩萨冠，与析木铁塔菩萨冠相同。该塔虽剥落严重，但可作为辽塔将佛化为菩萨像的一例。辽宁喀左大城子辽塔（精严禅寺塔）则于上下两层塔身全雕菩萨。

塔身主要表现菩萨形象，可能与辽代对菩萨的特殊崇敬有关。契丹原本信奉原始的萨满教，随着契丹王朝的建立及封建制度的发展，萨满教融入小乘佛教而成为辽王朝信奉的主要宗教。《辽史》卷四十九："太宗幸幽州大悲阁，迁白衣观音像，建庙木叶山，尊为家神。"《辽史》卷三十七："应天皇后梦神人金冠素服，执兵仗，貌甚丰美，异兽十二随之，中有黑兔跃入后怀，因而有娠，遂生太宗。"神人金冠成为辽代菩萨的典型形象。从艺术形象来看，析木铁塔砖雕菩萨形制特殊。衣装博大而宽厚，包被全身，腰腹间系宽松大带，具有中国北方游牧民族服装特点。菩萨的面目五官也是北方民族形象：高颧厚唇，与唐、五代及宋代中原地区菩萨造型明显不同。菩萨是梵语菩提萨埵的简称，原为释迦牟尼修行尚未成佛时的称号。菩提的意思为正，萨埵的意思为众生；言既能自觉本性，又能普度众生。罗汉修行精进便成菩萨，位次于佛（《释氏要览·三宝》）。在敦煌莫高窟造像中，唐五代时期的菩萨一律侍立于佛的左右，是面部有髭须的男子形象。在辽宁的金塔与广济寺塔，菩萨仍侍立佛侧，身形却大于佛近一倍，而菩萨神情之庄严，姿态之端丽，衣饰及背后佛光之造型都体现出自身的完整性，显示出明显的辽代突出菩萨的倾向。

3.须弥座砖雕中出现反映释、道并存或关系融合的题材

析木金塔须弥座西北面北端壸门旁，有组合立砌的两个肩上荷字之化生童子。一个肩荷"人"字，一个肩荷"我"字。佛教中的"我"是由梵文Atman意译而来，一般分"人我""法我"两种，简称人、法、我。佛教主张无我，但有的大乘派别又承认有一种超乎世间的"净我"或"大我"存在。道教认为修行者要专心做善事，来世可进升"三善趣"，即"人""阿修罗"和"天"。如果升入"三善趣"而仍未脱"轮回之苦"者，还需要断除"我执"才能超脱。所以在"人""我"的意义上可以认为佛、道两家是相通的。

辽代除佛教盛行之外，也流行道教。前已提及，辽太祖于神册三年（918年）五月乙亥诏建孔子庙、佛寺、道观，采取对释、道及孔子儒教兼容并包的态度。辽兴宗耶律宗真就是一位同时接受佛、道、儒三"教"的皇帝。《契丹国志》卷八记载他在夜宴时，和几个大臣也加入伶人乐队演奏，并命令他的后妃们改穿道士衣装。《辽史》卷二《太祖纪下》："秋八月丁酉，谒孔子庙，命皇后、皇太子分谒寺观。"辽上京临潢府便有天长观的建筑。在这个城市的街道上还有自由往来的僧尼、道士（《辽史》卷三十七《地理志·上京道》），说明道教在辽代是与释、儒并存的，道教的影响糅杂出现于佛教寺塔的艺术题材中就不足为怪了。在金塔须弥座砖雕人物中还有抱持酒瓮与背负酒瓮的酒仙及持羽状物的散仙。羽人、羽士本是道士的别称，酒仙的造型出现在佛与菩萨的座下也证明在辽代的佛教艺术中对于道教的兼容态度。而承袭唐、五代风格，造型严谨的朝阳云接寺塔，须弥座砖雕之背负瓮状物的人物形象却是沿袭唐代封建王朝的"胡奴"。析木金塔须弥座砖雕艺术中同时出现佛、道形象，同析木铁塔塔身每面雕一身菩萨一样，都是对来自中原的佛教雕塑传统形式的一种突破。

4.卧砖叠砌巨型组合砖雕开创了砖雕艺术的新纪元

以小印模印压制成的，有时是以多块卧砖联合拼接为巨幅连续花纹图案或人物故事情节的画像砖，和山东的画像石一样，早在东汉时期河南新野、四川成都以至辽宁盖县、旅顺一带已经出现。从考古发现看，在这些画像砖中已有凸出砖面的浮雕形式出现。辽代把这种手法用于塔身、须弥座的装饰，图像凸出更高于砖面，有半圆雕、高浮雕等形式。例如，北京天宁寺塔身大型佛、菩萨形象，浮雕技法纯熟，形象生动，是现存辽塔中之翘楚。但辽宁一带的辽塔，如析木镇铁塔等，须弥座各面的佛、菩萨、力士、狮子，以多层卧砖叠砌而成的巨型半圆雕、圆雕的砖雕形式，为中原地区少见。辽代将具有中原绘画效果的砖雕与辽代东北地区深厚、粗犷的石刻两种艺术风格相结合，创造了自己独具一格的砖雕艺术表现技法。这意味着当时的辽宁和东北、内蒙古等地区的建筑，在艺术形象和雕刻装饰设计方面技艺的发展。

5.砖雕用砖和砖雕工匠

东北地区民间至今俗称雕砖用砖为"金砖"。雕刻使用的砖在宋代称"坚砖"，所谓"黄閍（bēng）冈下得宝墨，古人烧砖坚于石"（宋人楼钥《钱清王千里得王大令保母砖刻为赋长句》诗），清代称"金砖"，"坚""金"音近。雕刻用砖质地细腻、坚实，在烧制时要经过把生泥过滤的程序，所以也称"滤浆砖"。辽代制砖工艺具体制造方法未见文献详载，但从近代制造雕刻用砖的方法，结合民间传说及烧砖老工匠的记述推知，大

致工艺程序可能是这样的：

A：选土，取地表下净、细、不杂砂石的"生土"（也叫"黎泥"）。宋《营造法式·窑作制度·瓦》："用细胶土不夹砂者。"

B：切细，生土日晒后切剁成细块，再曝晒至干燥为度。

C：研碎，将研碎之细粉用细筛筛净。

D：淘洗，将细粉投入淘洗水池，淘洗成为泥浆再经沉淀，取沉淀后细泥作坯土。

然后将坯土入砖模中制成砖坯，阴干至砖坯生霉为度，一般要3~6个月时间。然后入窑烧制，细火慢工也要费时2个月才能烧成。

辽代砖雕的工艺设计与建造者——砖雕工匠，未见有专门文献记载。而砖雕工艺与砖模的制作都需要专业的匠师。《辽史》卷五十《礼志·丧仪》关于丧葬制度有"官给工匠""造塔匠"的字样，应是包含雕砖工匠在内的统称。又辽王朝以战争手段俘虏奴隶，契丹贵族驱使各族奴隶从事各项农、冶、手工业的生产。《辽史》卷百十六《国语解》"应天皇后以太祖征讨、所俘人户有技艺者置之帐下，名属珊，盖比珊瑚之宝"，说明对这类人才的重视。这是挑选有各种生产技术、经验以及其他技艺者，其中建筑专业应该也包括在内。在神册三年（918年）建筑上京城，后来又为辽太祖阿保机营建陵墓、担任总工程师的康默记，就是阿保机侵蓟州时掠来的汉族人。

《辽史》卷五十一《礼志·军仪》关于对俘获人口中有技艺人士集中安置使用的记载不少。胡峤《陷北记》："遂至上京，所谓西楼也。西楼有邑屋市肆，交易无钱而用布。有绫锦诸工作，宦者、翰林、伎术、教坊、角觝、秀才、僧尼、道士等皆中国人，而并、汾、幽、蓟之人尤多。"明确记录了当时中原幽并（今河北山西一带）有技艺人不少在辽地营生。辽代遗存至今的塔、幢、墓志中又有关于"招募良工"的记载，可知砖塔雕塑的设计建造人员中还有招募佣聘制的自由民工匠。此外也有契丹内部罚没的奴隶艺匠。朝阳地区发现的乾统年间（1101~1110年）采石场摩崖石刻画上镌刻的作者姓名、创作时间及画面所表现的不加宗教色彩的民间生活内容，说明都是采石工人之即兴作品。由此可见，辽代砖塔雕塑艺术所具有的地方色彩与民族风格的形成，也饱含着土著艺匠的努力。

四　结语

本文列举的几座辽代砖塔所在地区，都是当时商业发达、经济繁荣、人口集中的重镇。云接寺塔所在地朝阳为辽太祖所建兴中府地；广济寺塔所在地锦州为辽锦州临海军；金塔、铁塔所在地析木镇本为辽东京辽阳府所辖析木县。析木县人口一千户，紧邻辽阳、辰州等"井邑骈列，最为冲会"城市，辽阳更是"市分南北，晨集南市，夕集北市"早晚集市不散的大城市。经济发达，宗教更为兴盛，众多辽代寺院、佛塔的建筑见于文献记载，而遗存于今的也很丰富。辽代佛塔砖雕艺术的创新发展，不仅仅有朝廷信仰的关系，社会经济的发达也是重要的原因。

辽代佛塔的砖雕艺术，从技巧至图样来自唐、五代以来中原匠人。更早一点，秦、汉的花纹砖和画像砖也有相当的影响。到了辽代，这种艺术在许多无名的各族匠师手中，经过深入的钻研默会，结合他们对社会活动的仔细观察，把所见各种形象，经过年深日久的揣摩和多番的创作，和佛教教义、佛学知识融通，使用简练、明快、真切的手法塑造了辽代自己风格的佛、菩萨、力士和供养人等形象。也使后世能够通过历史的遗留，在砖雕上见识到辽代的歌舞、伎乐等民俗情况以及当时民族的精神风貌。

契丹民族散居、游牧，习见自然风物，在造型艺术的发展上有强烈的追求写实的作风。在技法上追求生动自然、像真，使砖雕艺术在原来平面、浅画、浮雕的基础上，出现创造性的大型组装、卧砌半圆雕、圆雕等做法，丰富了砖雕艺术的表现能力。金塔力士仅以全脊连接塔身，挺然傲立，以青天白云为衬托，以大地山河为倚势的凛然气魄；须弥座下护法狮子张口、奋爪、探身欲出座外的见其形如闻其声的森然生怖的气势，都不是以平雕、一般的浮雕手法所能表达的。

大型砌砖人物、动物形象的砖雕组砌更是在整体的和谐之中富于多样变化，从制坯、烧造、打磨、组装，再不偏不倚地嵌置入塔身，才能完成预期的艺术效果。这不仅需要雕刻艺术造型手段，更需要一系列缜密的计算，一系列工序恰到好处的配合。这是何等浩繁、何等精密的创造性劳动。面对如此伟大的艺术遗存，谁还能怀疑辽代人民的智慧呢？辽代的砖塔和它的砖雕艺术永远屹立于民族艺术之林。在中国雕塑史的篇章上，这些光辉的艺术结晶可以永为古代契丹地方艺术所曾达到过的水平作证。

包恩梨
辽宁省博物馆

[本文是作者遗稿《辽代雕塑艺术》中的一章，原载《中国考古学会第六次年会论文集（1987）》（文物出版社，1990年），此次出版，据作者手稿与当时打印本，文字略有调整。徐秉琨整理]

THE SCULPTURE ART OF BRICK PAGODAS IN THE LIAO DYNASTY

Established predominantly by the northern Khitan Clan, the Liao is a dynasty rising after the Tang in Chinese history. Spanning two centuries (AD 907–1125), it coincided with the end of the Five Dynasties in the Central Plains and beginning of the Northern Song. Buddhism was the main religion, so numerous temples and pagodas were built in homage; thus leaving China a rich cultural heritage of intricate brick carving.

Here, let us briefly review the development of Chinese brick carving.

The birth of Chinese brick carving can be traced back to the Warring States Period, which means that the art of brick carving and the value of brick were produced almost at the same time. Related building elements such as tiles and tile-ends appeared simultaneously.

Early brick carving techniques tended to be rather monotonous. It mostly involved carving patterns onto unburnt brick with a tool or pressing with a mould and final firing in a kiln to create a portrait brick. By contrast, the production process for tile-end making was simpler. Although the brick portrait was rich in content, techniques were limited to line engraving or flat carving. Bass and high relief carving methods did not appear till the mid and end Eastern Han Dynasty.

The art of brick carving really came into its own during the Liao Dynasty. In the northeast where the empire was thriving, evidence of sculptures dating back to the Qin and Han Empires have been uncovered by archaeologists. Initial techniques were derived from traditional characteristic of intaglio and bass relief. According to *Ying-zao Fa-shi (Architecture Rules)* of the Song, the repertoire included Su Ping (Line-engraving), Ya-di Yin-qi (low relief), Jian-di Ping-sa (flat carving) and Ti-di Qi-tu (elevated relief or high relief). All portraits with these methods can be seen in the Liao temples and tombs. In addition, a most splendid, innovative technique of assembling colossal brick sculpture came onto the scene. This art form is brilliantly displayed in the Liao Pagoda architecture.

The Pagoda called Stupa in Sanskrit and Thupa in Pali originated from India and were used to house Buddha's sarira. Later this extended to worship and the storage of monks remains and holy scripture. The Pagoda pillar was initially a central column for supporting the top of cave temple and subsequently separated, to become an independent architectural form 'tower' out of mud and rock. In the development of Chinese pagodas, the stupa devoted to holy sacrament storage was confined to the finial and the main body expanded for viewing. In fact, brick pagodas did not exist till the Northern Wei Dynasty. A number of factors heralded the increased artistic freedom in brick sculpting, as opposed to stone. Firstly, the need to promote Buddhism using beautiful ornamentation arose. Secondly, with the benefit of conventional methods and the versatile nature of brick, images could be kneaded or moulded before firing in the kiln and later carved and polished on surfaces. In this way, brick carving gave Chinese Buddhist Art its distinctive flavour, as the various styles reflecting locality and period evolved.

Built in the first year of Zhengguang Era (AD 520) in the Northern Wei Dynasty, the Songyue Temple in Henan is the earliest dodecagonal tower with multi-eaves remained up to now. Using brick carving to decorate the body of the tower started a trend. Nevertheless, this early form did not deviate far from the original Western style. Most of them were decorated with plain surface, only the eaves, doors and windows were made of bricks. Furthermore, variety in shape was restricted and artistry was primitive. It wasn't until the Tang Dynasty with Buddhism thriving that both construction and engraving expertise reached new heights. An exquisite example is the Xiuding Pagoda, built during the Xiantong Era of the Tang (AD 860–874), at Anyang, Henan Province.

Certainly, temple building together with decorative brick carving reached a peak during the Liao Dynasty. The Liao Empire rising up in the North worshipped Buddhism. The founding emperor, Yelv A-bao-ji was already a follower before his ascension. In 902, as Yi-Li-Jin of the Yao-nian Hen-de-jin Khan Tribal Headquarter, he had erected a Buddhist temple at Longhuazhou (*The History of Liao, Vol 1: Biographical Sketches of Taizu Emperor · Upper Section*). Moreover, soon after being crowned in 918, a royal command was issued to build Confucian, Buddhist and Taoist temples. The reign of Yelv Zongzhen (AD1031–1055), Buddhism was valued highly, even generals, high ranking officials and the nobility would become monks and nuns, and live or worship in the temples (*Historical Records of the Khitan,* Vol.8). As a result, countless pagodas and Buddhist stone pillars started appearing in the Five Capitals. Folklore surrounding 'temples were built in the Tang and pagodas were built in the Liao' survives to this day.

The Liao pagoda's construction provided opportunity for the brick carving art to develop further. In terms of structure, the brick pagodas with multi-eaves tended to be built with an interior frame of imitation wood, which was fortified on the outside with brick sculpted figures of Buddha, Bodhisattva, disciples, and themes from religious folklore abound, including Buddhist donators and animal deities. As idolatry was an integral part of Khitan culture, it was a natural progression for construction of pagoda and adoration of Buddha's figure to merge, so that structure and external carving became inseparable. The shape of pagoda with its sculptural art became the prototype for successive dynasties: namely the Jin, Yuan, Ming and Qing. However, pagodas of the later presented only its lingering charm.

Built in the 10th year of the Liao Tianqing Era (AD 1120), the Tianningsi Pagoda in Beijing (Fig.1), 57.8 meters high, boasts a picture of Buddha-worship that images of Buddha, Bodhisattva presented a large scale assemblage style of typical Chinese painting. Bass-relief and modeled design are both used to great effect in portraying people and mounts meaderingly, and creates the impression of looking up at a long scroll.

Similarly, sculpted Buddha figures, inside niche-shaped doors on the sumeru podium of Jueshan Si

Fig.1
Pagoda body
Tianningsi Pagoda in Beijing
Photography: Jin Fengyi

Pagoda at Lingqiu in Shanxi Province, can be seen. Between the door and corner is a vivid and intricate carved strongman statue.

Built during the reign of Daozong (AD 1055–1101), the Wan-bu Hua-yan-jing Pagoda in Inner Mongolia is 45.18 meters high. It has carved figures of Bodhisattva, King of heaven and strongmen. The sumeru podium and brick walls between corbel brackets are carved vividly with Bao-xiang (designs of posite flowers) and peonies (Fig. 2).

Fig.2
Bao-xiang flower
sumeru podium of the Huayanjing
Pagoda

Built in the 18th year of the Liao Chongxi Era (AD 1049) at the Liao Zhongjing Site in Balin Youqi of Inner Mongolia, the Qingzhou White Pagoda is a seven-storied octagonal building with a height of 49.48 meters. Every story are inlaied with portrait bricks, and images of Buddha, Bodhisattva and strongman, together with scenes of musical dance and banquets are displayed distinctively.

The Daming Pagoda at the Liao Zhongjing Site of Ningcheng County is 74 meters high, with full faced Buddha statues carved in relief on the first level, showing the style of Tang architecture (Fig.3). The Liao Emperor Shengzong idolised the Tang ruler Li Longji so much that he changed his own name to 'Longxu'. Therefore it comes as no surprise that the Buddha on the body of Daming Pagoda is rich in Tang flavour.

The Yunjiesi Pagoda and North Pagoda in Chaoyang of Liaoning are all relatively early illustrations of Liao temple architecture, where the rich carving legacy of the Tang Dynasty can be seen. For instance, the facial expression and gestures of Buddhas, Bodhisattva, musicians and donators inside niche-shaped doors of sumeru podium echo the 'Zhou's Style' paintings of voluptuous females, considered beauties in the Tang. However, kids portrayed outside the niche-shaped door are the only images that show the features of Northerners. Further examples of Tang influence abound: the White Pagoda in Ji county, the Tianchengsi Pagoda in Panshan, the Yunjusi North Pagoda, the Jietaisi Pagoda, the Duobaofo Pagoda in Fangshan of Beijing, the Jingyan-chansi Pagoda in Kazuo and the Jiafusi Pagoda in Yi county of Liaoning and so on. Each possesses a unique style.

Liao Pagodas form a part of our nation's richest preserved architecture. Brick engraving continues Tang tradition whilst infusing characteristics of times, regional style and Khitan folk culture.

Fig.3
Buddha
Daming Pagoda

The Pagodas often stood in the 'saddle' area between two peaks. With a river flowing below plus sunrise from east illuminating the vital and magnificent sculptures, these grandiose pagodas remain testament to the best of ancient folk architecture. With the backdrop of Mother Nature, undoubtedly the Pagodas' amazing charm becomes one with the flowing river below and white floating clouds above. Stretched across the vast land and having withstood the baptism of war and the elements, preserved Liao Pagodas stand tall. Importantly, they make it possible for us to research and learn more about Khitan Empire's cultural achievement. Let us explore its unique features and artistic achievement by examining some examples which have stood the test of time.

1. Types and Technique of Brick Carving

From the position of inlaying brick-carving and the method of bricklaying, there appears to be three different arrangements.

(1) Vertical Brick-laying with Individual Brick: on the surface of the unfire brick, a brick is taken as an independent and complete picture. The entire image or pattern is imprinted with a mould before firing in the kiln. Then it is refined and placed facing outwards on the temple body or external walls of sumeru podium.

(2) Continuous Pattern Made of Connecting Bricks: a square brick sandwiched between two rectangular ones as a group, or 4 rectangular bricks side by side, create a single image or continuous pattern, patterns such as lotus petals, string of pearls, scrolled grass and cirrus are usually designed as border.

(3) Assemblage Figure Pattern: Raw bricks of various shapes are made first by using either small or big, square or circular moulds, according to need and style. Once after the kiln, the original partial portrait or pattern of each brick is put together for forming a complete pattern according to the pre-designed draft. Next, it is inlayed into the pagoda body, all sides of the sumeru podium and corners. This method is common with large sculpted figures of Buddha, Bodhisattva, Kings of heaven and mythical beasts, and style varies from deep to semi-relief. These ambitious projects were a true innovation for Liao craftsmen in blending architecture with the art of sculpting. The Tie Pagoda of hexagonal shape at Simu in Liaoning uses 28–30 layers of seven-meter thick bricks to lay Bodhisattva's sculpted figure. This makes the statues look like they are part of the pagoda. Such kind of practice is most commonly seen in Liaoning.

The technique used in the Han area of the Central Plains is slightly different. Patterns are often finished by bas relief and line engraving on a large plastered surface. The Liao pagodas exploited the full repertoire of bas relief, half relief, sculpture in the round and openwork style to create a three-dimensional effect. Emphasis was on conveying a sense of natural motion, activity, texture and space. Vivid illustrations are demonstrated in the Yunjiesi Pagoda's strongman in the four corners and the preserved headless figure on the northwest corner at the second level of the Simu Tie Pagoda. Whilst the trend for part of the Song Dynasty was following a rigid formula, the Liao brick carving was full of life and bursting with a distinctively mellow and timeless charisma.

II. Brief Introduction of Representative Liao Brick Pagodas

1. The Yunjiesi Pagoda

The Yunjie Temple, originally named Moyun is situated on the west side of Fenghuang Mountain, 15 km southeast of Chaoyang in Liaoning. Buit on a single-leveled sumeru podium, it is 37 meters high with 13 floors. The structure is squared in shape and multi-eaved. Mid-section of each side of the sumeru podium is an imitation wooden door, carved with 3 rows of 6 plum blossom shaped door-nails, beast-head shaped knocker holders and locks. Either side of this are three niche-shaped doors, covered with sculpted figures of Buddha, Bodhisattva, musicians, anupadaka (self-created) children and donators. Various floral, scrolled grass, cirrus patterns adorn the borders and corners. Respective images of deity, King of heaven and strongmen extending to all corners are exhibited in style of half relief and sculpture in the round. Each face abounds with bas and deep relief carvings of Buddha, Bodhisattva, flying Asparas and small divine pagodas. The upper part of Budda's figure and the divine pagoda is decorated with a ceremonial canopy and a pair of small flying Asparas. At the top of the divine pagoda are a pair of flying Asparas whilst there is no ceremonial canopy above Bodhisattva's head.

The four-sided sitting figures at the Yunjiesi Pagoda represents Four Buddhas of the 'Five Tathagatas' of Buddhist Tantra. Akshobhya in the east is accompanied by five cross-legged elephants whilst the southern Ratnasambhava by five cross-legged horses in the same position. Five cross-legged peacocks around the Amitabha is in the west and five garuda birds around Amoghasiddhi in the north. Each wreath-crowned Buddha statue have their feet on a lotus blossom and guarded by a standing Bodhisattva figure either side. Vairocan, or Vajradhra, is at the center of pagoda. The small divine pagoda has 13 stories with its body displaying an altar and a sitting Buddha figure in bas relief. Next to the divine pagoda are some inscribed Characters. On the left of the east is carved 'the Vimala-kirt Stupa', and to the right 'the Mahaparunirvana Temple'. On the south's left, 'the Mahabodhi Temple' and to the right 'the Maya Devi Temple'. Left of the west is inscribed 'the Jetavana Vihara' and on the right 'the Dharmrajika Stupa'. Finally, left of the northern face reads, 'the Monastery Venuvana Vihara' and on the right 'the Visahari Devi Temple'. The Bracket sets under the eaves are rather distinctive and reminiscent of the Huayan Temple in Datong.

The theme of 'Five Tathagatas', 'Eight Bodhisattvas', 'Divine Pagodas' are often used to illustrate Buddha's life. Indeed, these can easily be found, for instance, at the Wangtuzigou Pagoda below Fenghuang Mountain, in Chaoyang. Another example is the North Pagoda inside the city, first built during the Tang Dynasty and the exterior late restored during the Liao period, as uncovered by archaeologists. In terms of overall style and brick-carving technique, the Yunjiesi Pagoda is virtually a replica of the North Pagoda. However, the figure around the 2nd niche-shaped door on the southern side of sumeru podium, with thick lips, penetrating gaze and kneeling position, shows a typical image of ethnic groups of the Tang Dynasty. And at the right side of the 5th niche-shaped door is a male Khitan donator wearing a high hat and robe, with a Gu-duo on his hands (Fig. 4).

From the thriving Tang dynasty onwards, the practice in the Central Plains was to use brick to imitate wood. The tower surfaces tended to be both square and less decorative.

By contrast Buddhist pagodas built during the Liao and Jin Dynasties favoured the octagonal shape. Square towers had become rare. We can already see this transformation in the Yunjiesi Pagoda, a superb example of building on the Tang tradition whilst infused with the Liao distinctive style. Although square in shape, both main body and sumeru podium are full of intricately carved decoration. The Yunjiesi Pagoda truly exemplifies the early evolution of the art of Buddhist buildings and stands as a worthy milestone.

2. The Simu Tie Pagoda

The Tie Pagoda situated on the roadside of Liaoning's Haicheng no longer exists in its entirely original form. According to *the County Record of Haicheng*: 'Forty miles north west of the Simu city lies an ancient Tie Pagoda, over 30 feet high with seven levels and a temple at the bottom.' Mao Fengju, a local secondary school teacher in the 1960s, had ascertained its height of 23.6 meters by projection measurement. However the 1976 Haicheng earthquake shattered the peak and finial. In fact, Yao Rongguo, a long time resident near the tower recalls, before the quake, a sorghum was sprouting there and by autumn had grown as tall as the top. As these plants usually reach 3 meters, one can assume the finial is approximately that height. The locals fondly call this 'the little temple's roof top'. During the earthquake, a foot-high Buddha figure out of the pagoda crashed to the ground. The whole tower top smashed into 500 to 600 pieces and needed to be carted away. Sadly, both have been lost. Now the relic is only approximately 20 meters high.

The Simu Tie Pagoda is a hexagonal seven-leveled brick structure with multi-eaves. The base had to be fortified with cement in 1984 and lost its original look. The main body contains two floors. Underneath the first is a multi-petalled lotus blossom acting as a base, with a rather distinctive design of three petals supporting a bunch of brick carved Lotus stamen in the middle. At the same time, the giant blossom appears to be holding up the 5 meter-high first floor. A Bodhisattva figure carved in deep relief assembled style is inset into the tower wall each side. Part of a semi-relief ceremonial canopy slightly juts out above Bodhisattva's head and acts as a shelter. The statue is built with horizontal bricks of 28 layers. Each is 7 cm high so the total stature comes to around 2 meters. The background is made up of a pattern of 1 vertical (29 cm × 18 cm) and 3 horizontal bricks (20 cm × 7 cm). This combination is repeated at each corner to form the hexagonal undecorated pillar (the corner pillars are only half visible). Bodhisattva statue is depicted with a full face and square jaw. He is dressed majestically in wide sleeves, and the broad belt round the waist is so long that the ends brush his feet . Each foot rests on a lotus blossom with double-layered petals.

The Bodhisattva in the east is slightly inclined towards the southeast and has his clothing draped over the left shoulder whilst the right remains bare. A bracelet on the left wrist and an earring on the same side are clearly visible. Although it is unclear whether he has his palms together or grasping some object. However, his southeast counterpart is holding his palms together against his chest. He is dressed in the customary kasaya with wide knee-length sleeves and hem tassels brushing against his feet above the lotus blossoms. Two wide belts round the waist hangs down. Behind the holy figure, either side flutter sculpted strings of jade and pearl in bass relief, like horizontal flying objects. This classic look of earrings in both ears, hair tied back, sporting a crown adorned with carved cirrus and flowers of circular shape often seen on the Liao murals, is the simplest of all the six Bodhisattva statues (Fig.5).

The southwest figure is leaning towards the left, his bent right wrist clutching some pillar-like object, while the left sporting a bracelet is resting on his belt. Underneath the flower pinned to his necklace, a sash crosses over the left shoulder to the right and hangs down to the feet. He too is wearing an earring, visible in the right ear and with hair tied back, and the crown is decorated with three flowers. The west sided Bodhisattva's head faces right, showing his right hand and he is also wearing earrings, a necklace and the same hairstyle as well. The crown is ornamented with a single flower above the cirrus cloud patterns. In the left hand he holds a pair of small cups while the right is touching some object on them. The kasaya is draped over the right shoulder across to the left and down that knee. Simiarly, hem tassels brush the feet, resting on damaged lotus blossom carvings. The northwest counterpart also faces right, with hair tied back but only a flower decoration and no cirrus clouds on his crown. By contrast he is holding a pair of lamps in each hand and this time the kasaya is over his left shoulder and down to the knee. One brick-thick hem tassels touch his feet. In the background, a low relief halo is carved into the wall.

Last but not least, the northeastern Bodhisattva figure, like the others, has the same hairstyle. As with his northwestern counterpart, a single circular blossom adorns the crown. His right arm is lifted and the left is brushing against his waist cord. Like the previous statue, his monk's robe is draped over the right. The wide-sleeved kasaya is loose fitting with a belt round the front and the ends hanging over the knees. Similarly the hem tassels brush his feet. Not only is the northeast Buddha statue the best preserved but the shape of face and costume have the most distinctive Khitan characteristics (Fig.6).

The form of crown shows similarity to that worn by clay sculptured Bodhisattva at the Huayan Temple in Datong. Dress and adornment presents all typical features of the warm overcoat worn by the northern Khitan. It is very similar to the outer robe worn by a woman for burial in 1974 at Yemaotai in Liaoning and by wooden figures unearthed from the Liao Qing-Mausoleum. And also these decorative elements represent the process of sinicization and nationalization in the north since Buddhism was

Fig.5
Bodhisattva on the southeast side
Tie Pagoda in Simu

Fig.6
Bodhisattva on the northeast side
Tie Pagoda in Simu

introduced into China. The lack of a general bodhisattva of feminine tenderness, whole body looking like the image of the northern strongmen, are the characteristics of this pagoda's Buddhist figures.

A platform above the first-level's eaves supports its body of the second floor. The second is rather low, is approximately equal to the height of a wall between the first floor and the second. On each side sits a bass-relief carved Buddha with radiating aura. The figure is assembled using units of four square bricks end to end, each measuring 38 cm × 38 cm. The base occupies about a quarter of the statue's height. Tragically, only two bricks from the east-faced Buddha remain, one from the northeast, northwest and the southeast respectively. A rather pitiful decapitated strongman's statue barely 30 cm tall, is left on the southwest corner. Dressed in a long robe with a silk belt round the waist, hands crossed in front of his chest, he holds a lunge position. The raised wide sleeves behind the back is a classic symbol of the northern nomad horsemen's carefree spirit. Use of bass-relief to carve the belt gives the folds and silk texture a natural feel (A strongman figure is engraved on each corner column on the second level. All of them have kept their heads after restoration. Nevertheless, they are not what they used to be). Although no written records exist, the use of 6–10 groove-bricks of different depths on the second level appears to be a special Liao feature, along with the temples' shape and carving styles.

Tie Pagodas with two-layered-body is extremely rare. Even the Wooden Tower in Yingxian County of Shanxi has five levels. Its main body of the first is built with double eaves and a short wall separating it from the next; where a plaque can be hung. Hence, technically the temple has six tower-eaves. By contrast, though the Tie Pagoda is considered a multi-eaved structure, whether the bottom floor counts is still in doubt. Then the short tower-body of the second floor is a part of the multi-eaved construction. As far as its eaves is concerned, it is still a 'seven-leveled stupa'.

3. The Simu Jin Pagoda

Five kilometers northwest of the town of Simu, in Haicheng of Liaoning, stands the Simu Jin Pagoda, on a mound between twin peaks facing the Simu River.

The pagoda, 31.5 meters high, is a thirteen-floor octagonal structure facing north and south (the southern corner veers east 2 degrees). Its platform base was built on mountain rock with the technique of corbelling, where eighteen rows of bricks deeply keyed inside walls. Every side is decorated with six carved ganoderma-shaped cloud patterns.

The two-floor sumeru podium provides the foundation. Each level boasts a symbolical-style lotus blossom finished in semi-relief. And the whole pattern is made of the bricks which are moulded with lotus petals. Every slightly extended petal is delicately carved but bursting with energy. Both top tie-beam and bottom are adorned with strings of beads and curly cloud patterns either end. On each side of the sumeru-podium's midth is a Ying-xiang pillar, engraved with peonies while lotus leaves decorate the top and bottom. The sculpted patterns are only slightly elevated, as the highest point is levelled with the pillar's surface. This is in keeping with the Song's technique of Ti-di Qi-tu (elevated relief). Either side is a niche-shaped door, and outside it are ornamented with 48 assemblage style sculpted figures of musicians, dancers and servants. Although the faces and bodies are no longer intact, only those or the northwest, southwest, northeast, southeast and north are better preserved, what remain do serve to demonstrate the sophistication of the original art form. They are truly fine examples of the Liao brick architecture.

The various donators carved in animated form dressed in wide trousers tapered at the bottom, wearing necklace of jade and pearls and offering precious objects. Some figures are depicted as intoxicated, either sitting next to the huge wine urn, or holding the urn, or carrying the urn on its back. In particular, the four male dancers doing the Spin Dance, dressed in Khitan robes, long flapping sleeves and knee-high boots most represent the folk culture of the time. Either side of the northwest niche-shaped door stand a anupadaka child bearing two Chinese characters 'ren (human)' and 'wo (self)' on the middle of his shoulder, covering the entire surface. On the south-east side the bare shouldered flutist and female pipa player carved out of square bricks though damaged, still retain their elegant and lively pose. While the southwest wall has a player in Khitan costume, holding an instrument which resembles the horse head string instrument. It may very well have been a prototype. Despite the decrepit state of the performers' heads, the lively expression of them in action, either singing or recounting tales comes through clearly. Every corner of the sumeru podium is guarded by a strongman. They appear pumped up in a kneeling position, as if straining to hold up the structure. Although not well preserved, their lively spirit still shines through. By contrast, their counterparts at the southwest and northeast are more complete. They are portrayed in an animated form with mouths opened wide and necks strained, as if bearing a mighty weight.

Mid-way of this upper level is a wide niche-shaped door on each side; inside of which a lion stands guarding over the Buddha figure. Wearing a ferocious demeanour, stretching forwards on all paws with tensed muscles on the alert, the beast projects beyond the podium in the direction of the four points of the compass. Constructed of 10—11 layers of bricks, it has a width of around

80–90 cm and a height of 70–77 cm. Its majestic spirit is not diminished despite the objects inset into its eyes having been stolen. Similarly, on the outer walls of Li Jin's stone coffin from the Liao Dynasty (in the collections of Liaoning Provincial Museum), fierce looking 'tigers' and 'dragons' carved in semi-relief are found. This vivacious depiction of brick lions and stone tomb animals undoubtedly present realistic style of the northeast Khitan's decorative art.

On the upper-level podium, carved strongmen seem to be bearing the entire weight of the tower on their back. Virtually the whole figure is carved in the round. Against the contrast of the sky above, Their features appear spectacular. Sadly only the figure at the northeast corner is complete.

Above this upper sumeru podium level is a giant carved two-layered lotus blossom seat which appears to be holding up the main body. The whole tower stands 3.3 meters high on a base of 3.35 meters. Each octagonal face houses a niche with a semi-relief Buddha figure sitting on the sumeru podium. However the seat is so large that the front juts out. Outside, and Buddha is flanked by a Bodhisattva statue. The latter is made of bricks of 10–20 layers. The body is engraved in semi-relief while four heads sculpured in the round poke out of the main pagoda. Each is around 1.5 meters high, wearing crown and a necklace of jade and pearls, and with feet on the lotus blossom. The Bodhisattva at the east bears the Tang's full round face and square jaw style; but sadly both his and Buddha's heads are in poor condition. An engraved large ceremonial canopy hangs above the niche, with a smaller one over Bodhisattva's head, the top of which is decorated with a lotus.

At the top of Buddha's ceremonial canopy of each side are a pair of flying Asparas soaring into the clouds. They are in various poses, still or in flight, some clutching flowers and others with palms together. The south side flying Asparas is imitating the pose of a celestial maiden scattering petals, moderate in physique and appearance, with feet into clouds and long sash fluttering behind. As the masonry powder used for restoration was over thick, a lot of the original charm has been lost. The northwest flying Asparas rides above the clouds, making music and dancing. Whilst its eastern counterpart is stretching skywards above clouds. Similarly the northeast flying Asparas is flapping its arms, body twisted to the side with sash fluttering, making exaggerated dance moves, with half its body hidden by the clouds. The southwestern one holds a similar pose but it feels more natural and its head ornament retains the distinct Liao style. It is kept largely intact, although it was later restored. Whilst the northern flying Asparas is lying on its side in the clouds, with feet crossed, head supported by its arms, wearing a serene expression and appearing to be listening intently to the Buddha's teachings. Although these flying Asparas figures can't match the splendour of those at Dunhuang, they have their own lively, carefree and natural charm. The flying Asparas at the southeast and northeast are quite similar and should be made from the same drawings.

Tragically, due to the over-thick mortar used for restoration, the facial expressions and unique Liao style of Buddha and Bodhisattva figures have been lost forever. Plus the turbulent times which followed, it may well be that the Buddha figure and items in the niche are not original. The east Bodhisattva figure on the left is in poor state, exposing how the back was assembled with 15–16 layers of carved brick embedded into the body of the Pagoda. In addition, hardly anything is left of the statue on the right. The facial expression and pose of the north Bodhisattva statue display the fundamental principles of Liao brick carving whilst the strong jaw and round face clearly

Fig.7
Bodhisattva on the north side
Jin Pagoda in Simu

indicate the Tang's influence (Fig.7).

The Bodhisattva in the northwest is built with 18–20 layers of bricks, which is slightly different from the number of Bodhisattva in the left side of the east. This shows that in the pagoda, the brick-carved Bodhisattva's height is not exactly the same. The form of the Bodhisattva's crown is the same as those of the Huayan Temple in Datong.

To the north of the Simu Jin Pagoda, lies the remains of the Jinta Temple of the Ming, the tablet titled 'Construction of the Jinta Temple by Imperial Order' in the 3rd year of the Ming Wanli Emperor's Reign (AD 1575), the tablet titled 'the Record on Reconstruction of the Jinta Dabaolin Temple' in the 39th year of the Ming Wanli Emperor's Reign (AD 1611), the tablet titled 'the Record on Construction of Painted Jinta Temple' in the 41st of the Ming Wanli Emperor's Reign (AD 1613), and other three tablets seperately erected in the 43th and 58th year of the Qing Qianlong Emperor's Reign (AD 1778, AD 1793), and 22nd year of the Qing Jiaqing Emperor's Reign (AD 1817). The Memorial has all but disappeared. Only a pedestal in the shape of a tortoise remains on the original temple site.

Jin Pagoda Brick Specifications:
Two Types of Square Bricks: （1） 36.5 cm × 36.5 cm × 6.5 cm, 8 rows of grooves in the back,
　　　　　　　　　　　　　　 （2） 54 cm × 54 cm × 7 cm.
Three Types of Rectangular Bricks:
　　　　　　　　　　　　 （1） 23 cm × 39 cm × 7 cm, 2 rows of vertical grooves in the back
　　　　　　　　　　　　 with three crossshaped double grooves in the middle,
　　　　　　　　　　　　 （2） 24 cm × 41 cm × 8 cm, 9 rows of grooves in the back,
　　　　　　　　　　　　 （3） 18 cm × 39 cm × 7 cm, 6 rows of vertical grooves in the back.
The back of the bricks has a green tinge and due to the higher kiln temperature, a small number of slightly red ones have found their way into the Pagoda.

4. The Guangjisi Pagoda

According to *General Annals of Fengtian,* Vol.92, the Guangjisi Pagoda located in Jinzhou of Liaoning, was erected in the 3rd year of the Liao Qingning Era (AD 1057), and renovated in the Ming Dynasty. Archaeologists confirm its roots in Liao architecture.

It is a octagonal structure with multi-eaves of 13 levels and a height of 57 meters. The severely damaged single-level base measuring a width of 8.6 meters each side, had to be reinforced with smaller bricks in 1983.

There is only a sumeru podium of one-level remained in the pagoda. A brick carved Ying-xiang pillar each side is found mid-way up its body. The pillar partitions 5 kun-men (niche shaped) doors, inside of which sits a sculpted Buddha figure. The exterior sides are adorned with assemblage style engraved flowers in vases and devotees. The west facing carvings remain intact (Fig. 8) and the northwest's easily recognisable. However, those on the other faces have all eroded. The sumeru podium holds up a carved lotus blossom connecting with structure of bracket

Fig.8
Sumeru podium in the west
The Pagoda of Guangji Temple in Jinzhou

sets. It was supposed to hold a balustrade, which sadly have not withstood the test of time.

Each Pagoda face spans a width of 8.24 meters, separated by a corner column engraved in round relief. Each side, a Buddha sculpted in the same style resides on a lotus blossom in a niche, flanked by a semi-relief carved Bodhisattva statue twice the size. Spanning the top a ceremonial canopy sculpted in low relief can be found. Similarly, a carved canopy half the size lies above the two Bodhisattva figure heads. The right Bodhisattva statues on the southwest, southeast and east side are totally non-existent. The Buddha, Bodhisattva figures and ceremonial canopies are all shaped with carved flat-bricks. The Buddha statue facing the temple front wears a crown, with a halo reflected at the back. The others have their hair up in the shape of a screw. On the other hand, the crown worn by Bodhisattva figure is adorned with an engraved necklace of jade and pearls. The round aura behind his head on the wall is carved in low relief. Since the stucco has peeled over time, one can clearly see how the statue was constructed out of all the many bricks and grooves preset in the wall.

Above the ceremonial canopy sheltering the Buddha figure in the niche are a pair of Asparas in flight, at the very top, one on the right corner and the other on the left. Their form is no more than 2–3 leveled brick long and bear a striking resemblance to those behind the Buddha's halo at the Huayan Temple in Datong. By sharp contrast, the colossal Bodhisattva figure is over 3 meters high and requires 40 layers of brick. It is this huge difference in size that distinguishes this temple's sculpture from that of traditional Liao carving.

As part of Liaoning's coastal area, the humid climate in Jinzhou causes severe erosion. Consequently, the temple's original brick carvings and masonry are not well preserved. This allows us a glimpse into the process of how these colossal carvings were created. For example, we can see clearly the wooden support behind statues and where the Pagodas structural beams and mortise-tenon joints meet. Fortunately, as the carvings form an integral part of the building, except for an earthquake, they will not easily break off.

The grandiose depiction of Bodhisattva as solemn looking, round face and squared jaw, contrasts sharply with the Song Dynasty's meek and serene image. In actual fact it is closer to the full faced and rich representation of the Tang Empire and Five Dynasties. Jinzhou, established during the Liao Dynasty, bearing the same name to this day, located in the south of the northeast. And it was where Emperor Abaoji first used Han slaves for skilled labour. Geographically speaking, it provides a pivotal point for the Central Plains and northeast. Why this Pagodas carving style has a stronger Central Plains feel can be attributed to the geographical and national factors.

III. Unique Features and Themes

Multi-eaved pagodas of hexagonal and squared shape are rare, as the majority are octagonal. Often a sumeru podium is built on a flat base (or adorned with carved bricks that mainly imitates wooden brackets). On top of the seat, a budding lotus blossom appears to hold the main pagoda. The body contains one, two or even multiple levels. Each face is filled with carvings of Buddha, Bodhisattva, flying Asparas, Buddhist stories, egding patterns, as well as imitation wood doors and windows. Further up the tower, brick carving of imitation wood-effect buckets, which appear to be 'holding' the multieaves, can be seen. The peak is topped off with traditional pagoda finial. Therefore, we can see that the actual structure and brick carving craft are inseparable.

The Tang and Five Dynasties' legacy of style and shape is clearly seen in the structure of Liao pagodas and particularly the creative brick-carving on the sumeru podium and tower body. The predecessors' mould making tradition was a contributing factor. For example, the carved ornamentation on the Yunjiesi Pagoda, was modelled on the Square Pagoda of the Tang Xiangjisi in Xi-An, Shaanxi Province. In turn, the Yuan, Ming and Qing Dynasties inherited the octagonal shaped pagoda from the Liao era and the engraved adornment of base and body became the norm.

Nevertheless, the art of Liao pagoda's brick carving hasn't received the attention it deserves over the years. The four brick pagodas mentioned above gives us an idea of how the art of brick carving developed at an unprecedented rate during the Liao Dynasty. Following is a general analysis of some aspects of this art form:

1.Development of the sumeru podium

Originally the sumeru podium was a structure in its own right but evolved into the base with the arrival of Buddhist architectures. Its considerable surface not only enables the marvellous aspects of Liao brick carving to be showcased, but is also a distinctive feature.

Although carvings of Buddha, Bodhisattva and flying Asparas are deities in human form, once immortal they regain their own style. Since the Song Dynasty, portrayal of holy statues were subjected to rigid process and the same methods worn out by craftsmen. Not only did the quality of drawings limit innovation, facial expression and clothing had to be in strict keeping with the times and place. Moreover, sacred images left no room for spontaneous experimentation. However, the style of engraving on the sumeru podium demonstrates increased artistic freedom, as seen in the more varied and lively portrayal of the devotees bearing gifts. The representations give us a glimpse into the way of life and contemporary culture.

Without a shadow of doubt, the Khitan costume and moves of the dance figure amongst the donators, seen on the sumeru podium's southwest side of Simu Jin Pagoda, resemble the famous picture, *Zhuoxie Painting* (Zhuoxie means 'set up a tent to rest'), by the renowned contemporary artist in the Liao, Hu Gui. The dancers' clothing and steps are similar to those found on the Liao tomb murals. The sculpted figure's tap dance moves and spins look remarkably alike 'song-and-dance duet', 'yangko (a rural folk dance)' popular with Manchus and Mongolians in the northeast today. The way in which the stone carving brings to life the dancing scene with overlapping long sleeves, the angles of moves and how one naturally flows into the other, is truly amazing. It also provides a wealth of information on Khitan dance style. Could the Tang Whirl Dance have been like this? A further example is the demeanour, movement and costume exhibited by the 'opera figurines' on the Jin Dynasty tomb (*Cultural Relic*, 1979, 8) at Jiaozuo in Henan Province (Fig.9).

The novel craft of sumeru podium carving came into being from previous temple building designs and the practice of separating the pillar from the grotto art.

Fig.9
Figurine of musician unearthed
from the Jinyuan tomb in Jiaozuo of Henan
provided by Tao Liang

2. Unique characteristic of the prominent position of Bodhisattva on the pagoda's body

The adoption of a single Buddha and twin Bodhisattva figures as the main theme adorning the tower body is derived from Buddhist tradition. However, on occasion, the carving technique and arrangement style used, may make the Bodhisattva sculpture stand out.

Although Buddha's image appears on the more prominent upper part of Simu Tie Pagoda and the Bodhisattva figures below, the physical space above is actually narrower. Use of assemblage deep relief carving style has the effect of making the reclining Bodhisattva figures several times larger than the solo representation of Buddha. Although this practice remains rare, other examples do hint at the high regard with which Bodhisattva is held. Apart from the Yunjiesi Pagoda where Buddha's statue is bigger, the majority is smaller. Nevertheless, the octagonal multi-eaved pagoda known as 'the White Pagoda' is the exception, where Buddha's statue sits each side, wearing a Bodhisattva crown on the head. The style is similar to that found at the Simu Tie Pagoda. Despite the decrepit state of this pagoda, it provides a good illustration of how Buddha's image gradually blurred into that of Bodhisattva. Furthermore, the tower body at Kazuo (Pagoda of Jingyan Temple) in Liaoning, is fully covered with Bodhisattva sculpted images on the upper and lower levels.

This emphasis may be due to the avid adoration of Bodhisattva in the Liao period. Originally, the Khitan Tribe followed a primitive religion called Shamanism. But as both empire and feudal system developed, Shamanism became absorbed into Hinayana Buddhism and the latter took over as the predominant faith. *The History of Liao,* Vol.49 states: 'Taizong (emperor) moves the white-robed Guanyin (Avalokitesvara) statue from the Great Court at Youzhou to a purpose-built temple in the Muye Mountain. He adopts Guanyin (Avalokitesvara) as his family god.' *The History of Liao*, Vol.37 writes: 'The Yingtian Empress has a dream of a handsome man-god wearing a gold crown and holding a staff, followed by twelve beasts. Amongst them, a black rabbit leaps into the Empress's womb, resulting in the birth of Taizong.' Consequently, the gold-crowned 'man-god' became a classic symbol for Bodhisattva. From an artistic point of view, the style of Bodhisattva's carving on the Simu Jin Pagoda is rather unusual. The loose bulky clothing covering top to toe with wide sash round the middle, is reminiscent of that worn by Northern nomads. The stocky and thick lipped figures are in sharp contrast to the image presented in the Tang, Five Dynasties and Song Dynasty. In Sanskrit Bodhi is the abbreviation for Bodhisattva, a term referring to before Buddha became spiritually aware and attained enlightenment. 'Bodhi' means purity, and 'sattva' means all living beings. The word Bodhisattva denotes compassion, as once we gain awareness of our selfish nature and desires, we automatically want to be more charitable. Advanced arhat practice will ultimately lead to becoming Bodhisattva, second only in statue to Buddha. (See *An Overview of Buddhism: Three Jewels*). Carving statues found at Mogao Caves in Dunhuang, show the Tang and Five Dynasties Bodhisattva figure sporting a manly beard, flanking Buddha. This is replicated at both the Simu Jin Pagoda in Liaoning and Guanjisi Pagoda, with Bodhisattva's image is represented nearly twice the size of Buddha. Moreover, attention to detail of the solemn expression, graceful poise, clothing and halo behind, all point to the high regard Bodhisattva was held.

3. The coexistence of Buddhism and Taoism or the theme of fusion of relations reflected from brick carvings on the sumeru podium

Beside the niche-shaped door on the sumeru-podium's northwest side of the Simu Jin Pagoda, a single Chinese character is engraved on both shoulders of two anupadaka-child figures. One character reads 'Self' and the other 'Human'. The former is translated from the Sanskrit word 'Atman' and generally carry two meanings. One refers to being over-focused on the self and unable to let go of both ego and material possession. By contrast, the other is an enlightened self following Buddha's teachings. Buddhism advocates the absence of self, but some Mahayana sect concede the possibility of a purified altruistic self or great self able to transcend all worldly desire. Taoism teaches followers to devote themselves to acts of charity, in order to advance to the 'Three Good Paths', namely Deva, Human and Asura, in the next life. Those who succeed and yet unable to escape from the pain of repeated re-incarnation will need to work harder to let go of the ego. Thus, the concept of 'Self' and 'Human' in Buddhism and Taoism share the same meaning.

Both religion were popular during the Liao Dynasty. In the 3rd year of the Shence Era (AD 918), the Liao Taizu Emperor issued a royal command to construct Buddhist, Taoist and Confucian temple. Taoism, Buddhism and Confucianism sought to find common ground and co-existed peacefully. In fact, the Liao Xingzong Emperor Yelv Zongzhen followed all three religions. *Historical Records of the Khitan,* Vol.8 recounts how the Emperor ordered the imperial consorts to dance in Taoist robes while he and his ministers joined the orchestra in playing music. *The History of Liao: Main Events of the Liao Taizu's Reign* records, in August paying respects to the divinities, he orders the Queen and Prince to join him at a Confucian Temple and a Taoist Pagoda'.

Situated at Linhuangfu, in Liaoshangjing is Tianchang Abbey, a famous example of Taoist temple architecture. In this city's streets there are still Taoist monks and nuns walking freely (*The History of Liao,* Vol.37: *Geography · Shangjing Dao*). Let's make clear that Taoism in the Liao period co-existed well with Buddhism and Confucianism. Therefore it comes as no surprise that various Taoist influences appear on the creative themes of Buddhist temples and pagodas. Among the engraved figures on the Jin Pagoda's sumeru podium, are wine-immortals hugging and carrying urns on their back or divinities holding feather-like objects. Taoist priests were known as 'Feather Men'. The presence of wine immortal images underneath Buddha and Bodhisattva's seat is further proof of Buddhist art's tolerant attitude towards Taoism. Instead, the austere artistic style inherited from the Tang and Five Dynasties, dictate that images of Hu-slaves from the feudal system appear on the Yunjiesi Pagoda's walls. Both Buddhist and Taoist themed carving simultaneously appeared on the sumeru podium of Simu Jin Pagoda. These resemble the Bodhisattva figure on each face of the Simu Tie Pagoda and represent a breakthrough in the Buddhist engraving tradition from the Central Plains.

4. Giant composite brick carvings with technique of bricklaying heralded a new brick carving era

The practice of pressing with a small mould or sometimes putting together many square bricks to form a large scale continuous pattern or character story had started during the Eastern Han Period. It was widely used from Henan's Xinye to Sichuan's Chengdu, Liaoning's Gaixian and all the way to Lushun and surrounding areas. Archaeologists discovered that brick carving in relief had already existed in a primitive form. In the Liao dynasty, this technique was used creatively for the decoration of the tower body and sumeru podium. The image was higher than the surface of brick, and made mainly in two forms of semicircular and round carvings. For instance, the Tianningsi Pagoda's large scale Buddha and Bodhisattva figures are vividly portrayed, demonstrating these sophisticated carving skills. It is undeniably a crowning example of surviving Liao pagoda architecture. However, it is rare in the Central Plains to see on each side of the sumeru podium, huge Buddha, Bodhisattva, strongmen and lion statues, carved in semi-circular and in the round with multi-layered square bricks. Drawing techniques in this region, combined with a rough and ready stone carving tradition gave birth to this distinctive style. The artistic style and decorative engraving design give us a taste of architectural advances in the areas of Liaoning, the northeast and Inner Mongolia of that time.

5. Bricks for brick-carving and craftsmen

Even today, in the northeast, bricks used for carving are commonly known as 'Jin (gold) brick'. During the Song Dynasty, it is called 'Jian (hard) brick'. It is proved by a verse 'an inscribed brick-ink was found beneath the Huangbeng Mound, and the brick was burned as hard as stones' ('The Poem for Qianqing's Wang Qianli Getting the Bao Mu Brick with Wang Daling's Calligraphy' by Lou Yao). During the Qing Dynasty, the similar sounds 'Jin' and 'Jian' got confused, and carving bricks were called the latter by mistake. The texture of the brick needs to be both fine and solid. During firing, raw mud undergoes a filtering process, giving rise to the alternate name 'Filtered Brick'. Although no written record of the brick making process exists, we may surmise from contemporary production methods, folklore and snippets from old craftsmen the following procedure.

A: Soil selection: choose clean and fine mud without impurities.
B. Cutting: after being sun-dried, cut into small pieces and leave out again till all moisture is absorbed.
C. Crushing: sieve the finely crushed powder.
D. Panning: place the sieved powder into c panning pool till it becomes mud and falls to the bottom. This fine silt is now ready to be made into clay.

Next put the clay into the brick mould to make adobe. Then dry in the shade for around 3–6 months before slowly firing in the kiln on a low flame. This last stage will take two months till the brick is finally produced.

No written records exclusively of brick makers or craftsmen appear to exist. However there seems to be prescribed funeral arrangements for 'official-supplied artisans' and 'pagoda builders' in writing, according to *the History of Liao,* Vol.50, so presumably brick carvers are included. The Liao empire captured many slaves through war and put them to work in various fields such as agriculture, metal industry and manufacturing. *The History of Liao · Interpretation of Khitan's Characters* (Vol.116) states: 'By royal order of the Yingtian Empress and Taizu Emperor, the court seeks any artisans among the slaves with carving skills. They are as precious as coral'. This highlights the importance placed on artistic talent. Those with production skills, craftsmanship and experience, including architects were most likely to be headhunted. In the 3rd year of the Shence Era (AD 918), work began on the city of Shangjing and later the founding emperor, Yelv Abaoji's mausoleum. In fact, the chief engineer, Kang Moji, was a Han descendent who was made a slave when Abaoji invaded Jizhou.

There is plenty of Liao historical record of where a high concentration of indentured artisans were put to work. According to Hu Qiao's *Trapped in the North*: 'Went to the Xi-lou District of Shangjing, it was full of houses and shops that traded in cloth rather than currency. There were workshops for textiles, and eunuchs, Han-lin (in charge of Han-chinese documents), craftmen, Jiao-di players, musicians, monks, nuns, anc taoists who worked there were all Han Chinese. The Majority of them, however, came from the areas of Bing, Fen, You and Ji.' Surviving Liao pagodas, Buddhist stone pillars and grave epitaphs show efforts to 'recruit skilled workers', proving that amongst engravers and designers, there were free citizens for hire. On the other hand, those convicted of crime had to work for free as punishment. During the Qiantong Era (AD 1101–1110), dated and signed stone carvings of daily secular life were found in quarries, thought to be spontaneous art pieces created by the workers. The rich local flavour and folk culture demonstrated in the Liao pagoda carving can be attributed to the diligence of the engravers.

IV.Conclusion

Location of all above-mentioned pagodas was in bustling big towns with high density population, thriving businesses and strong economy. The Yunjiesi Pagoda is situated at Chaoyang, where the founding emperor first set up court. The Guangjisi Pagoda in Jinzhou was where the military was based. Simu, where both the Jin Pagoda and Tie Pagoda stand, was under the jurisdiction of the eastern capital, Liaoyang. A thousand households lived in Simu County, close to the most dynamic cities of Liaoyang and Chenzhou. Liaoyang was a large city with bustling markets from dawn to dusk and spread across north to south. A healthy economy and flourishing religion went hand in hand, leaving us a rich literary legacy of Liao Buddhist pagodas. Credit for innovative advances in the brick-carving art of Liao pagoda should go to Buddhism as the main religion, as well as the prosperity enjoyed.

The technique and sketching style of the Liao pagoda's brick engraving were handed down by artisans of the Tang and Five Dynasties, who had migrated from the Central Plains. Earlier still, Qin and Han brick-carved patterns and portraits exerted great influence too. In the Liao Dynasty, it was developed and created by many nameless craftmen after years of comprehension. Through careful observation of social activities, they have created various images that can be seen in their daily lives, and integrated them with Buddhist teachings and knowledge. Therefore figures of Buddha, Bodhisattva, strongmen and donators were engraved vividly by means of concise, lively and realistic techniques. Undoubtedly, these relics let the later generations to understand in depth the Liao's folk art and the national spirit at that time.

As the Khitan tribe tended to be nomads and exiles surrounded by nature, the artisans craved realism in their work. In pursuit of vital, natural and realistic representation, they transformed flat surfaces with dull colours into pioneering large scale compsite brick carvings in mid-relief and in the round. This breakthrough magnified the expressive powers of the brick carvings art. Exemplary illustrations include the back of the Jin Pagoda's strongman figure attached to the tower, standing tall and proud, with blue sky, white clouds behind, matched with the scene of vast land, mountain and river, adding to the overall grandeur. At the bottom of the sumeru podium stands a lion guarding Buddha, mouth wide open, claws out, ready to spring into action. We can almost hear its terrible roar from its majestic pose. All this cannot be conveyed by mere mural and plain relief.

These large scale brick-carvings with figures and animals are rich in harmony and variety. These must go through the whole process of blank-making, firing, polishing, assembly, and then being embedded in the tower-body, in order to achieve the expected artistic effect. This requires not only carving technology, but also precision calculation, and all the public order is just the right fit. What a vast and elaborate creation it is! In the face of such a great artistic heritage, who can doubt the wisdom of the Liao people? These magnificent artistic crystallisations form a chapter in carving history and forever bear testament to the exquisite standard achieved by the Ancient Khitan Empire.

Bao Enli
Liaoning Provincial Museum

(This article is a Chapter in *Sculpture Art of the Liao Dynasty*. It was first published in the 1990 in the '*Collected Theses from the 6th Annual Meeting of Chinese Archaeology Society (1987)*'. Reorganized by Xu Bingkun based on author's own hand manuscript with minor changes.)

辽代密檐砖塔简介

佛教起源于古印度，东汉时期传入中国，至南北朝时兴盛起来，得到了统治阶层的支持并在社会上受到普遍的欢迎。因为在此之前中国尚无一个成形的宗教能像佛教这样有理论体系、有宗教经典、有明确的教义、有大规模的信众。加之佛家的戒律与道德规范和中国传统的儒家教义有相通相合之处，如向善（劝人行善）、戒淫、戒杀（儒家反对滥杀）等，于是得以传播开来。佛家要求"三归（同皈）"（《魏书》卷一百一十四）即皈依佛（佛像）、法（佛经）、僧（僧人、传教者），亦即所谓佛家"三宝"，都需要建立固定的宗教建筑才能展现和容纳，这样，寺庙建筑便如雨后春笋出现在全国各地，与寺庙相联属的佛塔建筑也相应发展起来，只要是稍具一点规模的寺庙，寺内或寺外必然有佛塔存在，塔是寺庙建筑的一部分。岁久年深，寺庙或毁而不存，一些塔还巍然独在。

佛塔·密檐塔

佛塔和佛教一起也是从印度传入。印度的佛塔本是埋葬佛祖释迦牟尼舍利的一种佛教建筑，最初为纪念释迦牟尼，在佛出生、涅槃的地方都要建塔；随着佛教的传播，在盛行之地也多建佛塔，争相供奉佛舍利。传入中国的最早佛塔建筑是东汉时期的新疆"土塔"，都是建在佛寺之内，夯土版筑而成（据张驭寰《中国佛塔史》），塔身圆如馒头，下有基座，顶上出一细长的直梃，如其后的"塔刹"刹杆模样。山东嘉祥发现的一块画像石还刻有人物向这种佛塔膜拜的形象。江西庐山东晋时期的慧远法师（334~416年）墓塔也基本保留着这种"土塔"的形制。但由于中国文化传统影响力之强大，其后佛塔建筑的造型便脱离了"土塔"形象而趋于中国化，成为中国式的塔，超越了埋藏舍利的功能而成为一种标志性的宗教建筑物。

这些中国塔的平面造型有方形、六角形、八角形与十二角形多种，其形成的原因主要有两个：一是建筑技术方面受到中国传统建筑形制的影响，二是建筑理念方面受到佛经经文的影响。

第一个方面，中国式的佛塔建筑不可能不受到中国式的木构建筑的影响。所有的佛塔都或成为或仿自中国式的木构建筑。中国传统建筑是木结构，即以木质材料为框架再添填泥土、砖瓦而成。佛寺建筑也是这样，即使是石窟寺，窟的前面也有相应的木结构檐厦、廊柱、门窗。典型的木塔有北魏时期洛阳的永宁寺塔：为"九层浮屠一所，架木为之，举高九十丈"（《洛阳伽蓝记》）。可惜失火被毁，"火经三月不灭"。可能就因为木塔易燃而又塔高难救，故多用砖石材料造塔。存世木塔甚少，其著名者仅应县辽代佛宫寺大塔尚存。

用砖石建造的塔仍然是仿木建筑。尤其是砖塔，其形制、比例、细微之处都仿效得十分逼真。而从大的形制来说，中国的木构建筑都是矩形、方形，木材挺直，建筑物很难像以泥土为材料那样易于做成圆形，因而出现了这样的情况：开封嵩岳寺塔（北魏）为了接近圆形而做成了十二边形，其后一些佛塔又进一步做成了八边（八角，唐代已出现八角形单层塔）或六边（角）形。辽代佛塔中的八角、六角形，应即传袭自此。

第二个方面，佛经中曾明确说到十三层楼阁式方塔。《大般涅槃经·后分》（唐时僧人若那跋陀罗译）："佛告阿难：佛般涅槃（去世）荼毗（火化）既讫，……收取舍利置七宝瓶，当于拘尸那伽罗城内四衢道中起七宝塔，高十三层，上有相轮，……其塔四面面开一门，层层间次窗牖相当，安置宝瓶如来舍利。"同段经文还说到了十一层、四层及"无层级"即单层的方塔，彼此似有等级之分。云冈石窟寺（北魏）窟心多有方形塔柱的留置，有刻作三、四层者，应当与此有关。

而印度土塔传入中国之时，中国早已有了楼阁式高层建筑。汉墓出土的陶楼模型，其层高有达三、四层者。佛塔传入后，其建塔理念与中国的建筑技术相结合，于是出现了中国塔的一种重要的形制：楼阁式高塔。已烧毁的永宁寺塔就是九层楼阁式塔。而方形的造型又影响了有唐一代连同五代时期大量的塔式，辽宁朝阳一带多见方形辽塔，就是继承了唐代的传统。但楼层虽高，其高层主要是供眺望观赏之用，主要的佛堂建设、佛事活动还是在第一层，于是除因特殊需要而保持楼阁式者外，众多楼阁式塔渐渐压缩成为密檐式塔：第一层塔身之上的楼阁建筑压缩成一层层的矮壁，或保留示意性质的门窗或不保留，仅多层级的塔檐依旧存在，而这些檐下的斗栱柱枋很多都被取消，改变为简单的叠涩出檐，楼阁式塔改变成了密檐式塔。这种塔式在辽代得到了规模颇大的发展。而舍利塔要建十三层这一理念在一些辽塔中也普遍遵行。

密檐塔的建构

辽代密檐砖塔的情况复杂多样。以密檐层级而言，有十三层、十一层、九层、八层、七层、五层以至三层不等；以塔身装饰而论，燕云地区主要用假门假窗，传统辽地（即今东北、内蒙古一带）则主要饰以佛像。而这些以佛像为主要装饰内容的砖塔中，又有塔身每面皆出现佛像（暂称之"全佛式"塔身）和并非每面皆有佛像的不同。综合各方面情况来看，以全佛式塔身的密檐塔更具代表性，下面即以这类佛塔为主，介绍辽代密檐塔的基本情况：

（1）基本是砖构，间用少量木材。平面造型以八角形为多，也有方形和六角形。基本是南北方向。

（2）塔的整体结构，从下至上约可分为七段：

① 塔基。塔在地面建有台基，其上起塔。

② 塔基之下，即地下部分建有地宫。地宫的平面形状可作方形、八角形不等。

③ 须弥座。塔基之上为仿照佛座建造的须弥座。座的形状侧视如"工"字形，中为粗矮的束腰，上下两面扩展边缘于束腰之外。束腰的表面可有种种雕饰。须弥是大山的名称，以须弥山为座，说明佛的法力无边。以须弥座为塔座表示塔有佛的意义。须弥座可有一层或上下两层。

④ 莲花平座。须弥座上即为一层平面的仰莲平座（莲台）。座下有时有斗栱，即平座斗栱。莲花是佛教的基本花饰，源于佛本生故事：据说释迦牟尼诞生时，一手指天，一手指地，行七步，步步生莲花。在佛、菩萨坐立之处，往往衬以莲花。平座上有时又加勾栏。

⑤ 塔身。莲花平座上托塔身。塔身是密檐塔最主要的部分，楼阁式塔压缩为密檐塔之后，上面的层层塔身都变成了一段段的短壁，只保留最下层即第一层的塔身以供瞻仰朝拜。大多数佛塔的塔身只有一层，但也有的建有二层塔身，如喀左精严禅寺塔。

传统辽地塔身装饰的一般情况是，每面塔身的两边为两支立柱即角柱，塔身正中是佛龛，龛楣或作圆栱、眉栱、方栱不一。龛内为一尊坐佛，坐须弥座上，座的平面或方，或圆，或为多角形。个别塔身为立佛。龛外两侧，为立侍的两尊菩萨（或金刚），高处又有对飞的一对飞天。佛龛与菩萨头上，都有高悬的宝盖。整个设置如同一间佛殿，角柱即楹柱。有的塔身还出现"八大灵塔"（小灵塔）的雕塑。

这是传统辽地多数密檐砖塔塔身雕塑一种比较普遍的样式。但具体来说，又因塔形的不同而内容有所区别。如是方塔，以朝阳北塔为例，塔身佛像为密宗的"五方佛"（四面为四方佛，塔心室内置毗卢遮那佛即大日如来法身共为五方佛）。如是八角形塔，则八面塔身共现八佛。以阜新塔营子塔为例，佛顶宝盖之上有榜题佛号，经陈术石先生发现并考证，八佛佛号来源于《佛说八吉祥经》（南梁僧伽婆罗译）。而以锦州大广济寺塔为例，八佛中南面的主尊着菩萨装，结智拳印，仍是五方佛中的毗卢遮那佛。但另外七面的佛像却是"药师七佛"。1996年此塔维修，工作人员曾在北面佛尊的腿部发现一块砖，刻有"药师佛"等字样，因知此尊即药师佛。据此推论，主尊之外的七佛即应为"药师七佛"（见《药师琉璃光七佛本愿功德经》）。药师七佛还见于朝阳北塔地宫出土之经幢第三节石座的外壁雕刻，同刻有八大灵塔，榜题曰"八塔七佛名"（插图1）。就是说，广济寺塔塔身的八尊佛像为毗卢遮那佛和药师七佛。然则八角形密檐塔塔身虽皆有八佛，但八佛因塔而异，并不统一。如是六角形塔，以朝阳东平房塔为例，塔身的南面开券门，北面设假门，另外四面为四方佛，塔心中空为塔心室，推测室内原也应置毗卢遮那佛，即仍属于五方佛装饰。而绥中妙峰山小塔，六面塔身为五佛一菩萨，则又与"五方佛"无关。这都反映了辽塔装饰的多样性。

插图1
朝阳北塔石经幢上的八大灵塔和药师七佛图雕刻（拓片）

此外，还有些塔，其雕塑的主要内容不是佛而是菩萨，如海城铁塔，是为菩萨塔。

在各级密檐下一段段的短壁上，有时还贴壁饰有铜镜。

⑥ 塔檐。塔身之上即密檐塔的层层塔檐。塔檐的层数大都是奇数，常见者以十三层檐为多，偶数层只是极个别情况。檐层随着塔高而表现出一定的收分，使得整体的塔形看起来高大而稳定，比较美观。檐面排列瓦垄，有筒瓦、板瓦、瓦当、滴水等构件。每层塔檐周围有

多条屋脊（檐脊），脊下往往伸出一支木制的檐椽，悬挂一枚长体风铃，播送梵音。第一层塔身之上有檐下斗栱及槫、枋、椽、飞等构件，但由于塔身宽度有限，每面塔身只在两侧立有角柱，因而檐下的斗栱只有转角铺作和补间铺作两种，没有柱头铺作。

关于塔身的内部结构，需要说明的是塔心高层"天宫"的设置。塔基下多有地宫，其设置可能源于"舍利塔"，最迟在隋朝已经出现。《法苑珠林》卷四十引王邵《舍利感应记》："皇帝（隋文帝）皇后于京师法界寺造连基浮图以报旧愿，其下安置舍利。""连基浮图"应指连带有塔基构造的佛塔，"其下安置舍利"说明塔基下有地宫。但是塔上还有"天宫"的设置则少有提及。

而据近几十年的考古发现，几乎所有的密檐塔的塔心上层都还建有"天宫"。如朝阳北塔（方塔）塔心中空，被称为"塔心室"（插图2）。在第十二层檐的内心，塔心室顶部之上，有一个小小的方形宫室，长、宽分别为1.3、1.39米，壁高1.27米，有门道与外壁相通，是为"天宫"。宫的四壁立石板，有雕刻及题记。宫内置石函，内有木胎银棺、金舍利塔、鎏金银舍利塔、银菩提树、长柄石香炉、玻璃瓶、瓷瓶、题记铜板等物。又如沈阳塔湾无垢净光舍利塔（八角形塔），塔心中空被称为"腹宫"，圆形，上下可分为三室，三室之上仍是空心，可直通塔刹的顶龛。此"上室"应该也就是"天宫"，可惜早已塌落，遗物与中下层遗物混在一起，但其中有辽代遗物辽瓷等。

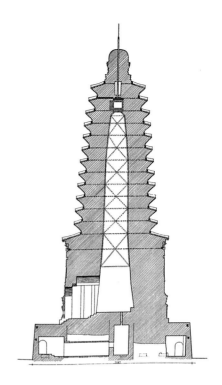

插图2
朝阳北塔塔心室

过去习惯上说密檐塔是"实心塔"，现在看来可能并不准确。除塔心高层建有"天宫"外，还有很多塔在塔心建有所谓"腹宫"即"中宫"。按"实"者，充满之谓也，与空相对。既有"宫"的设置，便是人为地留有空间，即使塔腔内有一段或多或少的填土，但从整个塔体而言，已算不得"实心"，故无须再强调塔的所谓"实心"问题。

⑦ 塔刹。辽代密檐塔的塔顶一般都随塔形而作成四角、六角或八角攒尖式顶，其上建有塔刹。刹座一般为受花或三花蕉叶式，有的上面又饰覆钵、仰莲等物。中间则立一高高的金属（铜或铁）刹杆，杆穿覆钵、圆光（宝珠）和多枚相轮等。刹尖外露，直刺天穹。

密檐塔的成就

建筑技术方面。中国传统的木结构建筑发展到五代北宋时期达到一个高峰，辽代寺塔也适在此时大量发展，辽塔建筑就为中国建筑史保存了一批重要的实例，从而大大丰富了人们对古代建筑的认识。这些实例出于辽代匠师们的精心制作，不仅仿木的构建和雕塑忠实严谨，而且做了很多创造性的发挥。辽塔高耸，不仅是佛教的标志性建筑，在当地也是最高的建筑物。宁城辽中京大塔高达74米（近于今天的三十层楼之高）且保存至今，在上千年前有如此建筑成就，实在是历史上的奇迹，也是世界建筑史上的奇迹。

在建筑和雕塑的设计方面，如塔座的设计。塔下建须弥座，可将塔身耸立起来，再加上一层或两层的莲花平座，使得塔身雕塑的佛教人物处于离地的高位，信士朝拜时需要仰视。而有的塔下又建造双层的须弥座，结合塔身雕塑的佛殿式设计，就更显得佛塔的尊崇与佛法的威严。这种座上加座的做法是创造性的设计，在庙堂建筑中没有先例。

还有塔身上小灵塔的出现。有些庙宇在佛殿内绘有佛传故事的壁画，以对礼佛者进行宣传教育。但在塔身狭小的面积内不可能有那样的表现空间，于是设计者雕塑了八大灵塔，这些灵塔代表着佛祖创建和传播佛教一生经历的八个重要事件和地点，是佛传故事的浓缩版，可使人们在礼佛时一样受到佛传故事的宣传教育。

在塔身雕塑的布局上，在下面的佛、菩萨与天上的飞天之间，有的塔用一层薄砖砌成一条横线，突出在塔身表面，将上下分割成两个画面，上面表示上空。其用意应是表示当佛在说法时，飞天在空中飞翔。这种表现空间的做法是一项精心的设计，在庙宇的佛殿建筑中难以做到。义县辽奉国寺大殿的做法是将飞天绘在屋梁上，但终不如让飞天直接飞翔在空中这样构思巧妙。

再有角柱的设计。在众多仿木的建筑构件中，柱头的阑额、普柏枋，檐下的各式斗栱及相关组合部件等都做得惟妙惟肖，一丝不苟，且各极其致，但毕竟皆受限于空间的狭小，雕塑匠师的创意很难有更多的发挥。唯有角柱与塔身等高，柱面也较宽，为匠师们提供了驰骋其意想的园地。庙宇的廊柱都是圆柱，而为避免千篇一律，有的塔上却出现有六角形角柱。有的砖塔更是在柱上雕出蟠龙，成为龙柱。还有的角柱上又雕出塔的形象、经幢的形象，成为塔柱、幢柱，这在庙宇建筑的木柱上都是做不到的。

雕塑艺术方面。辽代密檐塔上的雕塑过去很少有人提及，提到时虽认为水平很高，但也多一语带过。而如果真正面对这些雕塑作品，会让你眼前一亮，感到十分清新，生意盎然。其艺术成就之高，出乎人们的想象。

下面简单举几个例子。

以人物雕塑而言，塔身佛像两旁侍立的菩萨虽然饰有璎珞华绳，足踏莲花，但身材硕壮，整个气质是一些北方的彪形大汉，迥然不同于中原地区菩萨的柔弱恭谨。这表现了北方游牧民族的人物气质，同时也表现了契丹统治者尊崇菩萨的特点。

雕塑中飞天的身姿变化多端。或飞身向下，或侧卧空中如游泳状，或仰首，或支颐，或合十，或持物，头、颈、臂、足及双腿的动态表现潇洒多姿，衣袂飘举，加之流动的云彩衬托，充分展现了一种健康的女性之美。而飞天的特点原是"飞"，但有的飞天却不作飞翔态，而是端坐云端作谛听说法状，这种设计更是别出心裁。塔身佛殿的设置结合须弥座束腰众多供养人与乐师的活动场面，竟使人有一种做佛堂法事的感觉。这些浓厚的宗教生活气息，也给人们以深刻的印象。

在动物雕塑方面，如海城金塔须弥座探身狮子的威猛，其神态之逼真生动，肌肉解剖之合理，较之希腊罗马之著名雕塑作品也无逊色。而本是佛座下的护法狮子，有的塔雕竟表现为几对幼狮的嬉戏，一如人们常见的幼童或小动物打闹状，也匪夷所思。匠师们思想没有框框，尽管是严肃而神圣的宗教题材，在有关生活的内容上，其创作也随心所欲，驰骋他们的想象，为人们留下了一批可爱的艺术形象。植物雕塑中常见的仰莲平座周边莲瓣的丰硕饱满，生机蓬勃，也从视觉上带给人们以生活上的喜悦感。

辽代文化艺术的特点是"学唐"和"比宋"（历史学家陈述语）。辽圣宗为向慕唐玄宗李隆基，将自己的名字也取一个"隆"字叫作"隆绪"即耶律隆绪。辽道宗耶律弘基说："吾修文物，彬彬不异中华。"（《契丹国志》卷九）我们看到，辽中京大塔的正面佛像的形象，明显是模仿自龙门奉先寺主尊的唐雕如来。其他雕塑有时也能看到仿唐的迹象或唐雕的遗风。辽代塔雕继承了唐代雕塑的写实风格和大气，但又多了些野性和草原的生气以及一些北方民族鲜明的生活气息。

辽代匠师们高超的雕塑艺术水平集中反映在辽塔上，其与密檐砖塔高超的建筑技艺水平的统一和融合，奠定了辽代砖塔在中国建筑史和佛教建筑史上独特的地位。

　　（本文承张立先生、陈术石先生提供资料，谨此致谢）

徐秉琨
辽宁省博物馆

MULTI-EAVED BRICK PAGODAS OF THE LIAO DYNASTY

Originating from ancient India, Buddhism spread to China during the Eastern Han Dynasty and flourished in the Northern and Southern Dynasties. It was subsequently accepted and welcomed by both the ruling class at the top and grassroots at the bottom. Its popularity was partly due to the lack of any previous religion, which offered a theoretical basis, classical scriptures, clear moral code, sat well with Chinese traditional Confucian teachings. In particular, they shared common values such as inclining to goodness (encouraging acts of charity), controlling lust and not killing (Confucianism was against discriminate killing). Buddhist teachings preached 'Tisarana' (the Three Refuges), just taking refuges in the Buddha (the Buddhist statue), the Dharma (the Buddhist sutra), and the Sangha (monks), which are also known as 'Trirctana' (the Three Jewels) (*Historic Record of the Wei Dynasty,* Vol.114). However, for Buddhist teachings to manifest and be accommodated required concrete, permanent religious buildings. As a result, temples started sprouting across the land, along with pagodas which became an integral part. Any temple on a more modest scale would have a pagoda either inside or outside. With the passing of time, the former may have nearly disappeared but the latter stands majestically alone.

Buddhist Pagodas · Multi-Eaved Pagodas

The Buddhism and pagodas all spread to China from ancient India. Buddhist pagodas in India were originally a Buddhist architecture to bury Sakyamuni's sarira, and were built first in memory of Sakyamuni at the places where he was born and attained nirvana. And then with the spread of Buddhism, lots of pagodas were established in places where Buddhism was prevalent, and vied with each other for worshipping sarira. China's earliest example is the Eastern Han Dynasty's 'Tuta' (mud pagoda) found within the temple in Xinjiang. Made of rimmed earth slabs, the main body is round with a pedestal as foundation and a rod-shaped spire at the top (according to Zhang Yuhuan's *The History of Chinese Buddhist Pagodas*). The last appears to be a prototype of the pagoda finial to come. At Jiaxiang in Shandong, there is a stone mural with carved images of folks worshipping these types of pagodas. At Lushan in Jiangxi, the tomb tower of master Huiyuan (AD 334–416) in the Eastern Jin Dynasty basically retained the shape and structure of this 'Tuta'. However, due to the powerful influence of traditional Chinese culture, the shape of these structures soon evolved from the round 'Tuta' to look more like Chinese pagodas. It transcended the function of burying sarira and became an iconic religious building.

The planar shapes of these Chinese pagodas are square, hexagonal, octagonal and dodecagonal. And their formation is mainly caused by two aspects: one is the architectural technology influenced by traditional Chinese architectural forms and systems, another is the architectural concept influenced by Buddhist scriptures.

The first aspect. Chinese-style pagodas were inevitably affected by Chinese wooden architecture. All Chinese pagodas were to employ or imitate the architectural tradition of wooden buildings. The main structure would be a frame made of wood, with mud, and brick tiles added. Accordingly, the building of Buddhist temples followed the practice; even rock cave temples had eaves, colonnades, doors and windows all made of wood. One such classic wooden specimen is the Pagoda in the Yongning Temple at Luoyang, built during the Northern Wei Dynasty. 'A pagoda of nine floors with wooden frame, raising 90 Zhang (equal to 984 feet)' (*The Record of Luoyang Buddhist Temples*). Unfortunately it was destroyed by fire. 'The fire raged for three months'. Wooden pagodas burnt easily and its height meant putting the fire out difficult. As a result, the majority ended up being built with stone. Hence, surviving wooden pagodas are rare to be seen. The most famous example is the Large Pagoda in the Fogong Temple at Ying County, which was built during the Liao Dynasty.

Nevertheless, brick and stone pagodas still followed the practice of imitation wood architecture, especially the former and indeed in shape, proportion and detail, they all looked very realistic. In terms of large form, Chinese wooden structures were mostly rectangular and square. Timber by nature is dead straight, unlike earth, and cannot be manipulated into a round shape. Thus Chinese Buddhist pagodas were shaped consequently. The Pagoda in the Songyue Temple at Kaifeng, built during the Northern Wei Dynasty, tried to achieve a round shape by adopting a dodecagonal tower. This gradually evolved into octagonal (appeared in the Tang Dynasty) and hexagonal structures. Thus, came about the popular octagonal and hexagonal shape of the Liao Buddhist pagodas.

The second aspect. The Buddhist scriptures explicitly mentioned the thirteen-story square pagoda of pavilion-style. *Mahaparinirvana Sutra · Latest Version* (translated by the monk Jnanabhadra in the Tang Dynasty): 'Buddha tell the Ananda: Buddha was died and cremated,... put the sarira into the Qibao (seven treasures) Aquarius and built the Qibao Pagoda with a height of thirteen levels at Kusinagara... For placing the treasure bottle, the tower has a door and a window on each side.' The same passage also referred to the 11, 4, 'no level', or single-level square towers, which seem to be hierarchical with each other. In the Yungang Grottoes (the Northern Wei Dynasty), there are many indwelling square columns in the heart. And there are three or four levels carved, which should be related to this.

At the time when the India 'Tuta' started appearing in China, high-rise building of pavilion-style had already existed. A pottery model excavated at a Han Dynasty grave was up to 3 to 4 floors high. As the Buddhist pagoda was an iconic religious building, naturally it was also influenced by Chinese architectural preference. And hence an important structure appeared: the high-rise tower of pavilion-style. One such example was the burnt down 9-floored Pagoda in the Yongning Temple. The square-shaped structure also influenced plenty of towers in style during the Tang Dynasty and even in the Five Dynasties and Ten States Period. Towers of such kind are numerous in Chaoyang, Liaoning, indicating their inheritance from the tradition of Tang Dynasty. Though the towers are high, the upper floors are used mainly for viewing. Whilst the main hall for worship and religious activities are confined to the first floor. Therefore unless there was a special reason to preserve the high-rise pagodas of pavilion-style, they gradually shrunk to develop into the more compact multi-eaved model. Subsequently other floors apart from the ground one were compressed into layers of low walls, and doors and windows were either kept schematically or discarded. Although multi-leveled pagoda-eaves still existed, many structural components underneath eaves such as columns, brackets, crossbeams, and square beams had disappeared, deviating into a simpler design and technique of corbelling. Thus, the pagoda of pavilion-style evolved into the multi-eaved model. This change began during the Tang and Five Dynasties and took off during the Liao.

Structure of Multi-Eaved Pagodas

The situation of multi-eaved pagodas in the Liao Dynasty is complicated and varied. There are 13, 11, 9, 8, 7, 5, and 3 levels in the cornice hierarchy. As far as the tower body decorated is concerned, the pagodas in the areas of Yanyun are mainly decorated with false doors and windows, while in the traditional areas of Liao (Northeast and Inner Mongolia now) are mainly decorated with Buddhist statues. There is also a difference between these brick pagodas, some of them are decorated with Buddha statues on each side (so-called 'all-Buddha-style' pagoda) and others not. According to the comprehensive situation of all aspects, the multi-eaved pagoda covered with the all-Buddha statues decoration is more representative. Essentials of such sort of pagodas are as follows:

(1) They are brick structures with a small amount of wood. The flat shape is mostly octagonal, though there are hexagonal and square ones too. They basically face north-south.

(2) The pagoda's structure from bottom to top can be divided into 7 parts:

    a)   The Base: the base is constructed firstly and then the pagoda built on top.

    b)   The Underground Palace: beneath the base lies an 'underground palace', which is always square or octagonal.

c) The sumeru podium: above the base stands the sumeru podium, built to imitate the seat that Buddha sits on. It's in the shape of a Chinese character '工', with the middle squeezed inwards and the top and bottom spreading out. The surface of the tightened 'waist' might be filled with decorative carvings. 'Sumeru' means a tall mountain, to signify Buddha's endless power and the name also gives the podium a religious connotation. It is either one or two storied high.

d) The Lotus Flat-Seat: above the sumeru podium is a single level flat lotus seat. Sometimes, there are special decorative corbel brackets below the seat. The lotus is a basic Buddhist decorative pattern, originating from the Jakata tales. According to legend, when Siddhartha was born, one hand was pointing to the sky and the other down to the ground. After seven paces, each step grew a lotus. Thus, where Buddha or Bodhisattva sits or stands there is always a lotus flower. Occasionally the flat seat has Goulan (balustrade) added.

e) The Tower Body: the lotus flat-seat supports the body of the tower which is the most important part. After the pavilion-style towers evolved into multi-eaved pagodas, the upper floors became compressed into a series of low walls, leaving a single lower floor devoted to being worshiped. Mostly, the main body of the tower consists of one floor though some have two.

Generally speaking, the traditional decorations for the pagodas in the Liao Region are that two upright columns located at both sides of each body. In the middle is a niche, with its lintel in the form of a round, square, or brow-shaped arch. Inside the niche sits Buddha on top of the sumeru podium, the flat surface of which may be square, round or polygonal. Outside on either side, stands a sacred Bodhisattva figure or Buddha's warrior attendants. And there are a pair of Apsaras flying on high. In addition, ceremonial canopies hang above the niche and over the Bodhisattva statues' heads. The entire setting looks like a hall of Buddhist temple, the corner columns are the pillars. Some of the tower bodies also present in the sculpture of 'The Eight Great Pagodas' (The Small Divine Pagodas).

There is a very common form for the most traditional sculptures of the cornice brick pagodas in the Liao Region. However, to be more precise, the contents differ according to different shapes of pagodas. If it is shaped square, taking the Chaoyang North Pagoda as an example, the Buddhist statue on the body is the 'Five Dhyani Buddhas' of the Esoteric Buddhism (The Four Dhyani Buddhas locate at the four sides, in addition to the Vairocana Buddha, namely the Mahavairocana locating at the center, the five Buddhas altogether are called the Five Dhyani Buddhas). If is shaped octagonal, there are eight Buddhas at each side of the eight. Taking the Tayingzi Pagoda as an example, there are lists of Buddhist titles on the canopy on top of the Buddha, which was found and proved by Mr. Chen Shushi, the Buddhist titles originate from 'Astabuddhaka' (*The Sutra of Eightfold Auspiciousness*) (translated by Samghapala of the Liang Dynasty). If we take the example of the Great Guangjisi Pagoda in Jinzhou, the main statue of the eight Buddhas located at the south side wears the clothes of Bodhisattva, with the Vajra Mudra of Mahavairocana, which still should be the Vairacana Buddha of the Five Dhyani Buddhas. And the statues on the other seven sides are 'the Seven Past Medicine Buddhas'. This pagoda was renovated in 1996, the staffs once found a piece of brick at the leg of the Buddha statue in the north side, which was engraved with words of 'the Seven Past Medicine Buddhas', therefore, it has become well-known that this statue is the Medicine Buddha exactly. It indicates that the seven Buddhas except the main statue are 'the Seven Past Medicine Buddhas' (refers to *'The Sutra of the Vows of Medicine Buddha of Lapis Lazuli Crystal Radiance and Seven Past Buddhas'*). The Seven Past Buddhas have also been seen at the exterior sculpture of the third Buddha stone pillar found at the underground palace of the Chaoyang North Pagoda, which were engraved with the Eight Great Pagodas, in the name of 'The Eight Pagodas and Seven Buddhas' (Fig.1). In other words, the eight Buddhist statues on the body of the Guangjisi Pagoda are the Vairacana Buddha and the Seven Past Medicine Buddhas. Thus, we could know that although bodies of all octagonal cornice brick pagodas have eight Buddhas; however, they differ by each pagoda, and do not present in an uniformed shape. If it is a hexagonal pagoda, taking the Chaoyang Dongpingfang Pagoda as an example, the vaulted door opens at the south side of the tower body, and there sets a false one in the north, the Four Dhyani Buddhas locate at the four sides, the tower built hollow in its heart was functioned as a central room, suggesting that there should have set the Vairocana Buddha in the room, and it still belongs to the category of the decorations of the Five Dhyani Buddha. Moreover, regarding the Miaofengshan Pagoda in Suizhong, the six sides of the tower body bear five Buddhas and one Boddhisattva which is not related to the 'Five Dhyani Buddhas' at all. These all have reflected the variety of the pagoda's decorations in the region of Liao.

Fig.1
The sculpture of the Eight Divine Pagodas and the Seven Medicine Buddhas at Buddhist stone pillar

However there are deviations, such as the principal statue of the Haicheng Tie Pagoda being Bodhisattva and not Buddha.

Occasionally, bronze mirror may hang on the low walls, on each level under the multi-eaves.

f) The Eaves of Pagoda: above the tower body are the multi-layered eave tiles. The number of eave-layers is mostly odd, the common is 13, and even numbers are only rare cases. The higher the pagoda, the more narrow the roof since the aim is for a majestic stable shape, which is aesthetically pleasing. Rows of tiles on each eave-layer include cylindrical, plain tiles, tile-end and drip-end. Around every story of a pagoda is many rows of ridges with wooden eave rafters poking out underneath. At the end of the rafter hangs a long copper wind chimes for carrying Brahma sound. Above the pagoda body, is the corbel bracket, purlin, tie-beam rafter, and flying rafter components. Due to the tower body's limited width, each side can only accommodate two corner columns. Therefore there is just a bracket set on the corner and between the two pillars, but not on the columns themselves.

In terms of the internal structure of the tower body, the setting of 'Heavenly Palace' highly hanging at the pagoda center should be noted. Beneath the base lies an 'Underground Palace'. Its setting could originate from the 'stupa', which appeared the latest in the Sui Dynasty. *'A Forest of Pearls in the Garden of Dharma'* (Vol.40), quoted Wang Shao's *Sheli Guang Ji* (*Tractate on Response and Retribution of Sarira*): 'In order to complete the original wish, the Emperor (Wendi) and Empress built a stupa with subterranean structure at the Fajie Temple of Capital, to locate sarira stored underneath the stupa'. 'To locate sarira stored underneath' clearly indicates that there is a palace under the pagoda. However, that there is a 'Heavenly Palace' in the pagoda is rarely mentioned.

According to the archaeological discovery in recent decades, the upper level of the pagoda center for almost all cornice brick pagodas have built the 'Heavenly Palace'. For instance, the center of the Chaoyang North Pagoda (a square tower) is hollow, which is called the pagoda-hearted room (Fig. 2). At the heart of the 12th cornice, above the center room, there is a small square palace, which is 1.3–1.39 meters long, its wall is 1.27 meters high, the corridor is connected with the exterior wall, it is exactly the 'Heavenly Palace'. There are stone bricks at the walls of each side in the palace, which are engraved and inscribed. There are stone bricks inside the palace, holding a silver coffin with wooden body, a golden stupa, a gilding silver stupa, silver Bodhi trees, long-handled stone incense burners, glass bottles, porcelain bottles, inscribed copper plates, etc.. Another example could be the Wugou Jingguang (clean and pure) Pagoda (an octagonal tower) in Tawan, Shenyang, the center is hollow and called the 'Belly Palace', which is round and divided into three rooms both at the upper and lower sides, the space above the three rooms is also hollow and could directly connected to the top niche. This 'Upper Room' should also be the Heavenly Palace, however, it is such a pity that has caved in long before, the rest of it is mixed with remains in the middle and lower levels, and some of them are dated to the Liao Dynasty, such as the Liao ceramics.

The cornice brick Pagodas are used to being called the 'Solid Pagoda', which does not seem to be correct nowadays. Except the 'Heavenly Palace' built in higher levels, many pagodas have 'Middle Palace' in their lower level of the center. The word 'solid' here means 'fully filled', which is apposite to 'hollow'. As there is specific setting for the 'palace', enough space has been left intentionally, even though there could be several backfilled soil, more or less, inside the pagoda. However, in terms of the tower body as a whole, it can be called 'solid', therefore, there is no need to further emphasize on the so-called issue of the 'solidness'.

g) Pagoda Finial: the peaks of multi-eaved pagodas have always followed the overall shape of the structure to form quadrilateral, hexagonal or octagonal pyramidal roofs. Above them the

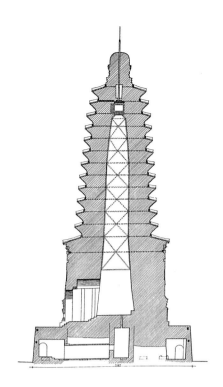

Fig.2
The pagoda–hearted room
North Pagoda in Chaoyang

pagoda finial can be found. Some are decorated with an inverted bowl or lotus. In the center stands an iron or a steel pole, which pierces the bowl, halo pearls and multiple Dharma wheels. The finial tip is exposed and stretches up to the sky.

Achievement of Multi-eaved Pagodas

In terms of architectural technology, traditional Chinese wooden architecture developed and reached a peak during the Five Dynasties and Northern Song Period. This was also a suitable time for the Liao pagodas and temples to grow in numbers. Consequently, as a number of concrete examples for the history of Chinese architecture, these Liao pagodas have afforded people of later generations great insight. They were made skillfully and detailedly by the Liao craftsmen. Not only was their imitation wood construction and carving skills loyal to tradition, but also their innovation in many ways. The Liao Pagoda is high, not only is it a Buddhist landmark, also the tallest building in the region. The Liao Zhongjing Pagoda in Ningcheng of Inner Mongolia is 74 meters high (equivalent to the height of a modern 30-story building), and still remains to this day. Such an architectural achievement a thousand years ago is a miracle in history, as well as in the architectural history of the world.

In terms of architectural and sculptured design, such as the design of pagodas, they created the sumeru podium, so that it would rise up with dignity. Then they added another level, the flat lotus seat of one layer or two, to enable the carved statue of Buddha to be seated off the ground, befitting his solemn and highly revered image. In this way, worshippers would have to look up to the sacred figure. Furthermore, some pagodas have two layered sumeru podiums with sculptures of Buddhist temple design, to emphasize the Buddhist pagoda's dignity and Dharma's majesty. Adding the flat lotus seat on top of the sumeru podium is truly an innovation in design and never been seen before in the history of pagoda building.

The Emergence of small divine pagodas on top of a pagoda is also a novelty. Some temples have murals with stories of Buddhism to promote the believer. However, it is impossible to find enough space in the narrow area of the tower body, so designers created the Eight Great Pagodas. This condenses the life story of how Buddha founded and spread the faith into eight milestones and spots. Whist followers are paying homage to Buddha, they are learning the founder's life history.

Among the carvings on the tower body, a thin layer of bricks juts out slightly to delineate two pictures. It separates the Buddha and Bodhisattva figures below and the flying Apsaras in the sky. Its intention is to show that at the same times as Buddha is teaching, the Apsaras are flying above. This clever demonstration of space is a meticulous design, and also difficult to be achieved in Buddhist temple building. They have attempted to create the same effect at the Fengguo Temple at Yi County, Liaoning, by having the Apsaras flying on the roof beam, but failed to match the same genius.

Regarding the design of the corner columns, amongst the components for imitation wood construction, the column head's architraves, Pubai Fang (a flattened square beam for supporting the corbel bracket) and various corbel brackets under the eaves and related parts, are all wonderfully and meticulously made. Nevertheless, due to the constraints of limited space, the sculptor's creative powers cannot be fully realized. Only the wider spaces of the corner columns and tower body can give free rein to the craftsmen's imagination. To avoid the sameness as round corridor columns of temples, hexagonal corner columns have appeared on some pagodas. Some brick pagodas even have coiled dragons carved on the columns. Yet others have sculptured the Buddhist stone pillar or the pagoda itself on the corner columns. This would have been an impossible feat on the wooden columns of previous temples.

In the past, few have commented on the Liao multi-eaved pagodas' carvings. If they did, they would acknowledge their high standard but only a passing remark. However, if you took the time to examine them closely, you would be pleasantly surprised by their refreshing liveliness. The standard of artistic achievements is simply remarkable.

Here are a few examples:

Compared to human sculpture, the Bodhisattva figures standing in attendance on either side are ornamented with Keyura (necklace of jade and pearls) and with their feet on lotuses. Yet they are built tall and sturdy, and their whole demeanor is that of robust men of the North. This is in sharp contrast to the timid and weak persona of their Central Plains counterparts. This shows the characteristics of northern nomads and the unique feature of the Khitan ruling class holding Bodhisattva in the highest esteem.

The poses of the carved flying Apsaras are varied whether swooping down, in a 'swimming' pose on their side, gazing up, resting their head in their arm, hands clasped or grasping some object, their head, arms, legs and feet's gestures are all elegant and changeful. With their floating clothing and the clouds as background, they fully exude a healthy sense of feminine beauty and charm. The characteristic of Apsaras is to fly, but some carvings don't seem to show them 'flying'. In some, they are simply sitting upright in the clouds listening intently to Buddha's teachings. This design is ingenious. Furthermore the lively scene of dedicated donors and musicians, assembled on the tightened 'waist' of the sumeru podium, is reminiscent of a Buddhist ceremony. This vivid atmosphere leaves a deep impression.

As for the animal engravings, for instance, the ferocity of the lion about to pounce on the Jin Pagoda's sumeru podium of Haicheng in Liaoning is portrayed in such on animated and realistic manner. The muscles carved all in the right places and with the correct dimension, and its excellence matches that of well renowned Roman and Greek sculptures. Yet, some carvings were supposed to be designed as guard lions, but are replaced by pairs of cubs frolicking around, just like young children or baby animals play fighting. It is really incredible! The sculptors do not seem constrained by the solemn, sacred subject matter. On the creative front, they have allowed their imagination to run wild and left us with some adorable artistic images. Equally lovely is the plant engraving of the full and rich lotus petals commonly used as the outer wall of the flat seat. They are pleasing to the eye and exude a sense of joy of life.

The Liao Dynasty's culture and art is classified as 'Learning from the Tang' and 'Competing with the Song' (a historian Chen Shu's statement). In order to show admiration for the Tang Emperor Xuanzong whose name was Li Longji, the Liao Emperor Shengzhong changed his own to 'Long', thereby becoming Yelv Longxu. The Emperor Daozong, Yelv Hongji proclaimed: 'Relics that I made as refined and courteous as that in Central China' (*The History of Khitan,* Vol.9). We can clearly see from the front facing Buddha statues in the Great Pagoda at the Liao Zhongjing Capital, that its inspiration we can detect an imitation of the Tang style or see its legacy. The Liao pagoda carving inherited spontaneity of its own and refreshing breath of the plains, together with a taste of the Northern folk way of life.

The Liao multi-eaved pagodas display an astounding standard in architectural mastery and artistic sculptural skills. This fusion of excellence guarantees Liao brick pagodas a unique statues not only in the history of Chinese architecture but also that of Buddhist structure.

(Thanks for Mr. Zhang Li and Mr. Chen Shushi to provide the information)

1 *The History of Chinese Buddhist Pagodas* 《中国佛塔史》
2 Yongning Temple 永宁寺
3 *The Record of Luoyang Buddhist Temples* 《洛阳伽蓝记》
4 *Mahaparinirvana Sutra · Latest Version* 《大般涅槃经 · 后分》
5 Jnanabhadra 若那跋陀罗
6 *Astabuddhaka (The Sutra of Eightfold Auspiciousness)*《佛说八吉祥经》
7 Samghapala 僧伽婆罗
8 *The Sutra of the Vows of Medicine Buddha of Lapis Lazuli Crystal Radiance and Seven Past Buddhas* 《药师琉璃光七佛本愿功德经》
9 *A Forest of Pearls in the Garden of Dharma* 《法苑珠林》
10 *Tractate on Response and Retribution of Sarira* 《舍利感应记》

Xu Bingkun
Liaoning Provincial Museum

寻找辽代文明，发现被忽略的砖塔雕塑艺术

　　为了给包恩梨先生留下的一部《辽代雕塑艺术》书稿补充图片，徐秉琨先生摘录了包先生曾在30年前实地考察辽代文明笔记中的描述，结合书稿上的内容编成词条，委托我按词条去搜集拍照。这就是此行的任务。

　　我开始做准备工作，包括拍照的人选、相机设备、交通工具、行程路线、导航仪、指南针等等。指南针是一个重要工具。不仅可用在行程上，还有一个重要的功能，是在拍摄现场要据以标注塔的方向。这也是考古工作最基本的要求。所以此次任务是一次目的性、专业性很强的工作，要在辽国疆域来寻找这些文明遗迹。辽代有上京（今内蒙古巴林左旗）、中京（今内蒙古宁城县）、西京（今山西省大同市）、南京（今北京市）、东京（今辽宁省辽阳市）五个京城，至今在一些市、地、县的博物馆和寺庙、岩洞、石窟、佛塔等处还有不少辽代的文化遗迹保存。这是一项细致的考古调查工作，如一条"词条"对拍摄要求的说明："名称：内蒙古巴林左旗林东前昭庙乾统九年（1109年）经幢，八面幢身、须弥座，莲座下岩石上的孔穴。备注：八面幢身七面刻陀罗尼经文，一面刻造幢人僧蕴崇等名字及建幢日期。须弥座各面刻龙、凤、牡丹、狮子、迦陵频伽等画面。莲座下岩石上每间隔130厘米有直径7、深2厘米作六角形分布的孔穴一共六个，估计是当时所建六角形的阁亭所遗留的柱础窝洞。"我是一个从事美术工作的专业人员，对于考古是一个外行。徐秉琨先生选择我去完成这项艰巨的任务，对我来讲既陌生又忐忑，又没有第二个人选。这对我是一次挑战，只能带着一颗好奇心去了。

　　既然接受了任务就要负起责任来，首先要选一位有牺牲精神、专业技术好、志愿与我同行的朋友。第一个想到了好朋友张守国先生，他有在西藏从事摄影工作二十年的经历，还是中国摄影家协会会员，比较专业，又有事业心。当我与他联系，说明为徐秉琨先生拍摄的情况，他立刻就定了下来，同时还跟我说开车去，什么时候行动，随时召唤。这样一切准备工作完成就可以出发了。

　　我们在2016年7月开始了时空穿越的里程，走进了辽代。

　　一上路就遇到了很多的问题，没有想到的事太多了。导航仪不好用，到了山上不知道从哪里进山，进了山那苦就开始了。有些时候，不是车拉人，而是人要下来推车，没有其他帮手，只能靠我们自己。还有时车开到山上，路断了，没有了路，只有山民们采蘑菇的小道；当我们车开到尽头时，前面是山谷。像这样的地方一般都在古栈道上才能接近拍摄目标，但是需要徒步，我们还要带上二十多千克的设备行走，像石窟、神庙，还有些佛塔都是在这样的环境里进行拍摄，非常危险。辽代佛塔最不好拍，尤其是山上的塔，因它的拍摄空间有限，底座空间最多也就两米宽。而大部分佛塔又建在山顶或山腰，为了拍全就得往山下退，经常是一个人在后面抓着前面的人才能工作，一不小心滚上一两个跟头也是常有的事。有时胳膊、手、腿都是伤，为了工作，回到住处连衣服都不敢脱，因为血已经把衣服粘上了，几天后才好。汽车轮胎爆裂、补胎、修车，这些我

们都经历了。我们每天寻找的工作都是处在一个思想高度集中、紧张状态下进行的，寻找辽代文明变成了冒险的里程。

当我们的镜头触碰辽代文明的那一瞬间，一切的艰难、辛苦都烟消云散了，我们就想多拍一些。而拍摄又是一个特殊的工作，与时间、天气、光线、角度都有直接的关系，有时要等上几天才能工作。目的就是要把辽代文明遗迹更好地搜集回来，完成徐秉琨先生、包恩梨先生交给我们的重任。

在寻找辽代文明的过程中，我们还搜集到了二十余座辽代砖塔雕塑资料。在拍摄的过程中发现了被忽略的辽代雕塑艺术。我是一个专业的美术工作者，学习美术史时只知道秦汉、魏晋南北朝、隋唐、五代、宋元明清，辽便没有了，从古代佛教雕塑艺术看，有著名的敦煌莫高窟、麦积山石窟、龙门石窟、云冈石窟等等。这些不学美术的人都知道，我想这是中国的文化学者们在这方面工作，做出的杰出贡献。而对北方少数民族文化关注很少。但是辽代在中国的大地上建朝立国有两百多年的光荣历史，并且北方的马背民族也曾经影响过世界。西方世界史学者说，是中国的马背民族创造的马镫，使欧洲提前进入了现代社会。

在辽上京（今内蒙古巴林左旗），辽祖州城址也是此次拍摄工作中一个意外的收获。我们拍完巴林左旗博物馆的藏品，馆长王未想先生说还有点时间带我们去辽祖州城看看，说那山上有座公元4世纪的"石房子"。到那儿时天色已晚，登上山看到"石房子"，我被眼前这一幕惊呆了（插图1、2）。这是一件极其现代的雕塑作品，空间、环境太美了！震撼！正像雕塑家康斯坦丁·布朗库西（Constantin Brancusi，20世纪最具影响力的雕塑家，被誉为现代主义雕塑先驱）说的那样，"建筑是有人居住的雕塑"。从房子里往外望，一个美丽的画面呈现在眼前，辽阔的草原、起伏的山脉，在晚霞的辉映下极为壮观，很有诗意。"石房子"的背面，有巍峨壮丽的大山映衬，此时"石房子"有了思想

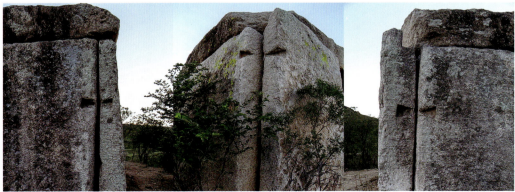

插图1
石房子
建筑正面与墙图腾
约建于公元4世纪

插图2
石房子
建筑房角结构
约建于公元4世纪
巴林左旗辽祖州城址

和性格。这用七块巨石板构成的建筑，就是奇迹。仔细观察每一块石板的结构，凹槽和斜面巧妙地组合在一起。这种结构具有很高的科技含量，对化解草原飓风、抗击自然灾害起到了重要的作用。大门造型独特，雕刻出了契丹族的性格，大开大合的气派。室内有一块长4.3、宽2.5米的巨石板平放在地上，博物馆专家称其为"床"。石板东面刻有九宫格图案，由东向西有两条刻出去的线条指向中段，石板中段也有不同形状的符号，西面也有格子图案，看上去很像古代的占卜图，与天文应有一定的关系，对草原出行、族长们议事等起到很重要的作用。总之这些使建筑更具神秘的色彩。在北面墙与屋顶之间留有一道缝隙，和南面的门相对应，起到了很好的室内通风作用。外墙下刻有网格图案，当围着作品欣赏的时候，发现建筑南面房顶的侧面刻有图案或文字。虽然已经看不太清楚了，但图形的结构还可以辨认。墙上还刻有人形的图腾。我把看到的图腾和博物馆馆长王未想先生讲，他说这是第一次听有人说上面还有图腾，他没有看到。他说"不可能，我在这工作几十年，到此无数次，也没有听说过图腾的事"。我指给他看，他看到了图腾，很兴奋，说"这是新的发现"。我向他讲述了我的经验，根据房子建造独特，透露那么多的信息，不像普通的建筑，我想也不像考古专家们说的"石室"或"石房子"那么简单，准确地应该叫"神庙"或"神殿"。所以我带着疑问和好奇认真搜索建筑上每一个细节，发现了图腾。另外要想获得这些信息，首先要具备一定特殊条件，特定的时间、光线、天气，最重要的是发现者心里得有才能看到。这是我多次考察寻找古代岩画的经验，我们今天能够看到就是这些条件都具备了。而我们的到访不知是运气还是感动了上天，让我们了解这么多的故事，或许它们太需要与现代人交流了，希望把它们的故事传达给更多人了解它们的曾经。在辽阔的草原上它是契丹人用智慧创造出的伟大不朽之作。对研究古代文化，今天的建筑学、力学、环境科学、民俗学、美学都有很高的学术和实际应用价值，尤其是对现代艺术更有启发借鉴意义。

巴林左旗博物馆有很多北方草原民族的雕塑，其中一件契丹族人物的石雕让我震惊（插图3）。属圆雕，有一人高，头及一只胳膊缺失，但这都不能抵挡它的光芒，雕塑大气、浑厚，人物形态精准、流畅、生动、概括、简练中不缺少细节，在服装上微微地刻画出装束的特征，使石头材质发挥到极致，一只胳膊无，剩下一只手与另一只手叠合。在考古学家们眼里此件雕塑是缺憾，而在艺术家们眼里是完美，残缺使雕塑产生了一种神秘的艺术效果，更有想象空间，启发艺术的再创作。作品反映契丹艺术家对石头材质的领悟，不是把石头模仿成一个人，而是用一块有生命的石头创造出永恒的艺术。我想现代雕塑大师亨利·摩尔（Henry Moore）若看见这件伟大的雕塑也一定会感慨的，因为它代表了现代思想和原始的精神。我们搜集的辽代雕塑，尤其是辽砖塔雕塑，鲜明地展现了辽代雕塑艺术发展的全过程，也是一部立体的、形象的、生动真实的表现了辽代文化艺术发展的思想史。

辽宁省海城析木金塔是拍摄雕塑的重点，按词条提示我们找到了金塔，词条上的内容有："序号38.辽金塔塔身每面一龛，龛内坐佛，龛外左右立侍菩萨各一身，菩萨以10~20层卧砖装砌，半圆雕全身，以4层卧砖圆雕头部探出塔身。""序号40.辽宁金塔塔身北面菩萨像，佛与菩萨顶上都有组装卧砌砖雕宝盖，每面宝盖上有组装卧砌飞天一对……作奉献、舞蹈、弹奏状……""序号35.辽金塔须弥座瘿项柱西侧壶门内外组合砖雕伎乐人、舞者、使者，共计48身……人物形象有结裤管佩璎珞捧献者、跪坐调筝者、怀抱马头琴弹唱者、持卷者、策杖者、持羽者，抱、负酒瓮者，弹琵琶者、吹笛者、着契丹装作旋风舞者，以及肩'人''我'字化生童子者……"这是由徐秉琨先生从包恩梨先生30年前的实地考察笔记中归纳的部分对金塔的描述。我们接近金塔时首先看到的是雄伟、浑厚，带有欧式罗马时期风格的建筑，八面雄壮的圆柱结构的塔身在山林绿色衬托下，有阳光的照耀，看上去像它的名字一样——金塔。真漂亮！当我们来到塔的近处时惊呆了，词条上的内容，就剩下飞天、护法狮子还算比较完整，其他均遭到破坏，佛像一尊都没有。所有的菩萨头部和部分手都没有了，须弥座上的48身像没有一个完整。我把这些情况汇报给徐秉琨先生，同时把刚拍到的图片传给徐先生，当时徐先

插图3
翁仲
石雕
巴林左旗辽祖陵神道
2003年清理

生一声长叹"……晚了，晚了……"我又把拍到剩下的几尊护法狮子照片传给徐先生看，徐先生很激动连说："太好了！太好了！""这在中国绝对是没有见过。""这肌肉块的表达像罗马雕塑，太有张力了。"让我一定好好拍一拍，随后我又给徐先生传了几张飞天的资料，徐先生又一次激动地说："这我以前怎么没看到，还有这样美的飞天，很像唐代造型。"同样让我多拍一些，并注意雕塑的制作工艺。接下来我们拍摄工作就好开展了，不用按词条上的内容去找着拍，而是自由地把能拍的、有艺术价值的全拍下来。最不好拍的就是北面塔身上的那对飞天，比较高，地面后退没有空间，下面是山谷。为了更好体现雕塑的完美性，我们做出了冒险的行动，我先退到山边手抓住树枝，看一看还有没有可能再发展一下的空间。我把一切探索好了，让伙伴张守国先生在我的前面把脚架支上，我用身体挡住他，拍完后一看很成功。这对飞天是金塔飞天里最美的一对，从造型上看有唐代的遗风，明显受唐代影响，体态上肥胖、饱满、舒展、唯美，头部很像唐三彩中的女俑，表现手法上，写实、概括、奔放、生动，属半圆雕塑，写意、飘逸自在，在中国佛教飞天雕塑中属上等佳作。

金塔须弥座上的护法狮子有几尊保存得还算完整，雕塑震撼、写实、大胆、饱满、肌肉夸张、彰显力量，头发表现装饰性流动的线条非常生动，面部表情带有笑意，很有幽默感，仿佛在对来者说托起这塔不算什么，在动作设计上几尊护法狮子各不相同，头部是从须弥座里向外探出，生动的样子很灵活，有的是一只胳膊就能托起佛塔，有的是用一边肩膀就能扛起佛塔等等。与面部表情统一在一起，一个是轻松，一个是力量，肌肉块在这里的表达象征着力量，面部的笑意象征着善良。从整体上看护法狮子很亲切，它既是对真善美的一种表达，也有另一种含义，在邪恶面前将使用法力使罪恶粉身碎骨，看起来更世俗化，用现代语言形容就是更接地气。辽代艺术家用写实的方法、浪漫的思想解读了宗教的思想，创造出辽代草原民族独特的文明。包括塔须弥座上的力士与护法狮子的手法一致，夸张的肌肉，在动作设计上，有用肩膀扛的、脖子顶的、手托的，面部也是带有笑意，很生动。在塑造上彰显了艺术家们极高的艺术审美，演绎着雕塑们的激情。我们从残缺的一个力士头部看，塑造得很含蓄、概括、奔放、生动，将人的精神塑造得栩栩如生（插图4）。雕塑中头部质感的泥巴味道使我联想到法国著名雕塑家罗丹（Rodin）塑造的巴尔扎克头部的感觉（插图5），虽一个是亚洲10~12世纪的创造，另一个是欧洲19世纪的创造。但它们之间太像了，仿佛就是一个人的作品。还有塔身上残缺菩萨的手更生动地展现了艺术家们的雕塑创作激情及熟练的技巧，仿佛能欣赏到艺术家创作雕塑的过程，上泥、拍打、手一拧一抹的动作，畅快地将泥巴味道演绎得淋漓尽致、生动传神。从塔上留下的残缺雕塑作品看，充分地展示了辽代艺术家们高超的技术和艺术审美。它与过去朝代的佛教雕塑有很大的不同，创作材料、地点、空间、审美都不同。材料选择了用原始自然纯朴的

插图4
力士
须弥座砖雕
辽宁海城析木金塔

插图5
巴尔扎克（局部）青铜像
1893~1897年

土加水混合的泥巴，这也是雕塑家们常作为创作稿使用的材料，先用泥巴创作出艺术家们所要的形象，再由艺术家们按其需要的雕塑语言选材，比如：有喜欢青铜材料的，有喜欢石头材料的，还有喜欢木头材料的，等等。泥稿就是艺术家创作的试验过程，它可以随着艺术家们的想法而创作，也可以随意地修改，使艺术家们真正获得自由变化，当完全达到了艺术家们所要的理想效果时，稿就结束了。然后再用艺术家需要的材料转化出来，是放大或缩小，还是石头、木头或翻制青铜材料等等，一件雕塑作品才算真正完成。所以泥稿是雕塑作品前期创作阶段，不同的是辽代艺术家们直接用泥塑造出形象再转化成建筑的格式同土坯一起放到砖窑里烧，烧成后再砌到塔上，这样一件作品初步完成，恰恰是辽代艺术家选择这种方式来创作雕塑，减去了一个再复制的环节，使雕塑原创的激情被保留了下来。在金塔残缺的被脱去保护层的雕塑作品中可以看到雕塑的技术，比如：上泥用木板拍打、刻、雕、挖、手抹、搓各种形状的泥条，绳形、球形等，直接使用在雕塑上，自由地被艺术家所运用，轻松地将个性思想释放出来。这些方法对创造出辽代独特风格的雕塑艺术起到了重要的作用。而金塔的雕塑艺术最大的特点就是体现在"塑"字里，凸显出辽代雕塑家们艺术的审美修养和个性。在金塔雕塑上还有一种材料是不可忽视的，那就是建筑使用的砌砖的黏合汁。用石灰粉加糯米粉混合制成的黏合汁，相当于古罗马人发明的水泥，辽代的草原民族把建筑用的石灰、糯米粉进行更加细致的研磨，经过不同的比例配制把它抹在塔上，而雕塑的表现面也要涂抹。但是它不是简单的涂抹，而是由艺术家们进行的第二次创作，把在烧制搬运过程中的损伤用它"找"回来。这时艺术家们在创作过程中的塑痕发挥了重要作用，它起到挂住石灰与糯米粉混合泥浆的作用，又起到了修复填补雕塑不足的作用，同时也起到给雕塑抛光的作用，使作品更坚固、饱满、生动完美。这种材料经历了上千年的检验，依然完美（插图6）。同时我们发现在辽代因砖塔所处的地域不同，材料的产地也不一样。虽然都是石灰和糯米粉混合而成的黏合汁，但产生的雕塑表面质地效果是不一样的。有的像玉石雕刻，有的像陶上挂釉，还有的像大理石雕刻，这样丰富的表面效果在辽代砖塔雕塑里都能看到。辽代草原民族创造的方法，在今天的意大利继续演绎着。世界著名意大利雕塑家、艺术家布鲁诺·瓦尔波特（Bruno Walpoth）被称为"本世纪最伟大的木雕艺术家之一"，他的木雕作品就有这种技术，在雕塑过程和最后完成时在作品表面都要涂抹涂料（插图7、8）。在创作中反复涂抹，这个过程是修整的过程，也是抛光的过程，它既保留了雕塑过程的痕迹，又有圆润艺术作品视觉效果的作用，同时也创造了他独特的艺术风格。而辽代砖塔雕塑艺术中，这种技术被广泛使用。

　　金塔的雕塑艺术反映出契丹族的智慧，同时也发展了佛教雕塑艺术。在辽代之前，我们看到的石窟、寺庙的石刻或泥塑，都属于室内雕塑。辽代砖塔雕塑则是在室外，材料、地点空间都与前朝不同，艺术审美也不同。雕塑作品凸显生动、奔放、随性的草原民族的特征。也可以这样说，当佛教雕塑从寺庙、石窟走到室外之际，也是雕塑艺术发展之时，

插图6
天王
塔身砖雕
辽宁喀左精严禅寺塔

插图7
女孩（一）
当代木雕
布鲁诺·瓦尔波特（意大利）

插图8
女孩（二）
当代木雕
布鲁诺·瓦尔波特（意大利）

好像辽代艺术家们在这一刻的思想被解放出来。北方草原民族的性格和气质，自由、豪放、自信、勇敢、善想象、善创造、对生活充满热情，在金塔雕塑里都得以体现，开创了中国古代室外雕塑艺术的辉煌。

当把寻找到的二十余座辽砖塔雕塑的资料搜集在一起时，可以看到装饰性雕塑在辽代得到了空前的发展。从一个主题浮雕飞天看，飞天浮雕表现形式极其丰富、水平甚高。种类有半圆雕、高浮雕、浅浮雕。手法有写实的、有装饰的、有用线的，还有用色的。这不但是中国辽代装饰性浮雕艺术大全，也是中国历代装饰性雕塑之最。当我兴奋地观赏飞天时，仿佛是欣赏辽代艺术家们在探讨艺术美和运动规律。例如：锦州广济寺砖塔的塔身上的一对飞天浮雕构图和造型与传统的飞天有所不同，飞天造型呈现三角形状构图，被一个由云朵编织构成的巨大弧线所包围，产生了强烈的震动感，使飞天有一种向前滑翔的力量，构图气势磅礴，这一切是艺术家通过构图创造出来的视觉效果。从飞天形象上看缺少了以往的佛性，而更多地接近现实中的人，看上去更亲切，脸部表情更祥和，发式造型有了现代感，身体塑造出像人一样的温度、生动、概括，仿佛是在创造一个新生命。

在雕塑语言上，从胸到胯再到膝盖这三点之间的塑造和压缩、高低节奏上控制得非常完美，使飞天塑造的气度、薄厚、量的变化都恰到好处。艺术家巧妙地运用装饰线将飞天与云朵连成一个整体，这不仅没有破坏飞天的主体，反而加强了飞天在空间自由、自在的滑翔运动姿态。艺术家在飞天形态上很注重细节的刻画。如：两条腿形态变化节奏上，向上的腿压在下面腿上，下面小腿的脚搭在上面小腿上，产生了形体的扭动，给三角形状赋予了动的变化，强调了飞天的优美、形态舒展。尤其是飞天身上的披肩、披巾、裳裙、璎珞、天衣、带饰上线的变化，雕刻完美。它的宽度、密度、深浅、层次、弧线、直线、线的长短运用都达到了很高的境界，因此说线语言就是这对飞天雕塑语言的灵魂，展现了辽代艺术家们敢于创造的精神，将人的思想带入宗教里。这对滑翔式的飞天就是很好的证明，艺术家们创造的灵感应该来自北方草原上的生活。这对飞天特别形象地表达出草原上空中的雄鹰展翅翱翔，雄健、优美、自由自在的精神。滑翔式飞天的独创，也象征着草原民族进取、勇敢地追求美好、自由的理想，同时也彰显了辽代艺术家们敢于创造的智慧和丰富的想象力。如今再来欣赏这种象征辽代北方草原民族精神的飞天，虽然人类已经实现了在天空中飞翔的梦想，像世界极限运动"翼装飞行"在空中的滑翔。但是人类追求自由自在美好的理想从未停止过，辽代艺术家们创造的艺术和思想在我们这个时代继续演进着。

锦州广济寺砖塔雕塑飞天融合了草原民族的精神和审美，在这座塔上每一件雕塑都能看到这种精神的闪耀。如：塔身上的菩萨更典型，每尊菩萨塑造出各自的特征，从西面一对菩萨看，雄浑、奔放、生动，仿佛是在塑造民族英雄英俊、潇洒、慈祥、智慧，给人一种亲切感。与传统的菩萨雕塑形象大不相同，根本看不出是菩萨，只有身上穿的袈裟能辨认出是菩萨。从雕塑看，以往传统规范、唯美的平静没有了，一个崭新的形象被创造出来。尤其是雕塑语言上，强烈的刀法技术凸显了雕塑家们的自信与艺术观点。近似疯狂的线刻，刀功的演绎，使砖雕塑艺术达到极高的境界。在雕塑菩萨头部上看雕刻刀功运用，压、刻、拍、削等非常自如，刀法干净、犀利，形象塑造结实，质感很像石雕，概括、生动、饱满，使砖雕塑的泥巴味道转变成刀功的世界。菩萨身上袈裟刀法更加豪放、自由飘逸、线性的表现，狂放到几乎找不到一根直线，艺术家的激情在这一刻全部被释放出来，好像雕塑家们在享受创造过程的快乐，使人浮想联翩，似乎都能感受到艺术家在雕塑中嘴里哼着草原小曲、抖动双肩、跳着契丹人独创的"胜利马步舞"在向英雄致敬！看着这充满激情的线刻，真令人陶醉。很显然这生动的线刻来自艺术家观察现实获得的灵感，它不再是传统的模式。虽然在雕塑里找不到一根直线，但是艺术家却创造出一个真实鲜活的形象，那些飘逸的曲线表达了客观真实的存在。可以想象艺术家是在塑造现实中一个真实的英雄站在高处迎着微风，服装与身体接触啪啦啦拍打的视觉效果，反映出辽代艺术家们对客观真实观察精细，尤其在细节上更注重生命形态的刻画。比如：两腿的膝盖微微向里一扣，关键点的捕捉准确地将从头到脚的定力表达得稳如泰山，再伴随着袈裟、飘带生动的刻线，使整个形体英姿挺拔精神。这

插图9
迪亚哥半身像
1950年铜制雕塑
阿尔贝托·贾科梅蒂（瑞士）

插图10
低浮雕
1932年青铜
马利诺·马利尼（意大利）

插图11
菩萨
塔身砖雕
辽宁锦州广济寺砖塔

不仅是艺术家们用现实表现手法创造的新形象，更重要的是将中国传统的宗教雕塑艺术推向新时代，真正地进入了现实主义创作雕塑阶段。这个时期各民族的文明相互融合，各种风格雕塑艺术蓬勃兴起。同时也是传统雕塑与现代雕塑的分水岭，这是辽代砖塔雕塑艺术所做出的贡献。艺术家们勇敢地强调艺术的本体，用刀法探讨艺术的美学思想，创造出独特艺术风格的雕塑艺术作品，像这样用刀法语言挑战传统是任何朝代都无法比拟的。如今这种艺术形式深受现代雕塑家的青睐。20世纪存在主义雕塑大师、瑞士的阿尔贝托·贾科梅蒂（Alberto Giacometti），20世纪超现实主义雕塑巨匠、意大利雕塑家马利诺·马利尼（Marino Marini），他们都是用刀法解释艺术问题，并且有着独到的见解，为世界雕塑艺术发展做出了巨大的贡献（插图9、1C）。而中国10~12世纪辽代雕塑家们更早地发现用刀法表现艺术的审美（插图11）。更加伟大的是艺术家们用刀的功力刻出了中国最早的现实主义雕塑艺术，也可以说是开创了中国雕塑史的新纪元。这是和艺术家们所处在一个伟大时代分不开的。在辽塔上的很多雕塑作品里都可以见证这段历史的辉煌。

辽塔雕塑艺术用自己独特的民族精神解释了宗教的思想，也创造出草原民族独具特色的文化。一路上搜集到的辽塔上的雕塑艺术，处处渗透着草原文化的印迹。在辽宁朝阳北塔、凤凰山云接寺方塔、朝阳王秀子沟大宝塔等，塔身上佛像的莲花座下都有马的形象出现。还有辽宁喀左精严禅寺塔须弥座上，在修行者的故事里也有马的形象。马代表北方马背民族的精神，它也象征着草原民族对马的热爱和崇拜。同时也是草原民族献给宗教最珍贵的礼物。

在辽塔上还有许多反映草原民族特征的人或物，如：北京房山石经山上云居寺北塔须弥座上人物身上佩戴的饰物、用皮革制成的服装等，都是艺术家们用精湛的技艺、写真的手法记录的辽代北方草原民族现实生活的特征，也是艺术家为宗教文化做出的创造和历史的记录。

从辽塔雕塑艺术还可以了解到草原民族开放的胸怀，他们很注重学习、吸纳、交流、传承不同民族的先进思想和技术。

很多塔上的雕塑都有唐代特征，如：辽宁海城金塔上的雕塑风格和内容就有唐代的特征，在须弥座上还有反映使者身穿唐代的官服等。在朝阳八棱观塔上的雕塑和塔的设计是反映唐代文化的代表。从须弥座的雕塑内容上看，如同反映唐代宫廷生活的连环画，有侍从、乐人、舞者等身穿唐服，包括乐器、用具造型构图，画面刻画得栩栩如生，很可惜全部遭到严重的破坏，现在只能从残留的痕迹辨认出图形，不然我们可以看到辽代人眼里的大唐生活，更可以享受到辽代艺术家们雕塑艺术的风采。这个塔的建筑比较特别，佛龛和佛像是裸露在塔身的，莲花座呈大半圆形结构，佛像腿呈圆雕形式，造像具有唐代风格。残留的一尊佛像服饰及莲花座上的莲花设计都很精细，疏密安排非常完美，佛像身穿的袈裟、服饰随着形体的变化而设计，衣纹已经不存在了，只有形体和规范的图案，整体形象浑厚、饱满、概括、唯美，是典型的装饰雕塑，这也是比较少见的质量很高、艺术性很强的具有唐代艺术风格特征的雕塑作品。很遗憾因被严重破坏，佛头部和手都缺失。但从中仍能看出辽代对唐朝文化的崇拜和敬畏的意识。

在内蒙古巴林左旗博物馆里藏有辽上京南塔上的雕塑，有佛像、菩萨、飞天，还有带翅膀的飞天，有道教里的道士，这些都是塔上的原作。在这一座佛塔上融合了这么多的文化和思想，可以看出当时的辽上京文化有多么的繁荣昌盛。在庆州白塔也发现了带有翅膀的飞天，来自不同民族的使者、供养人和西方使者。朝阳凤凰山云接寺方塔须弥座南面雕有西方的使者，特征非常鲜明，卷发、长胡须形象。北京房山石经山云居寺北塔须弥座上带翅膀的天使、欧亚的使者、乐人等，这都表明辽代对外文化交流的开放、包容、进取的思想。而这一切是艺术家们用雕塑艺术做出最直观、真实的形象记录，对今天了解、研究辽代文化思想发展有着重要的现实意义和历史价值。

在辽宁喀左精严禅寺塔的雕塑中，看到了受欧洲文明影响特征的雕塑，如：菩萨形体塑造结构上，有明显受到古希腊时期雕塑影响的痕迹。在塔身一层西南面上一对菩萨雕塑得非常完美，突破了以往的复杂，而是简练，把衣纹归纳成几条有节奏的直线，舒展、清晰，很有建筑感，这种审美的表达在中国艺术雕塑史上从没有呈现过。这是辽代文明与欧洲文明碰撞、吸纳，创造出来的新雕塑语言的表现方法，尤其是在写实方面上更能显现出艺术家超强的功夫，在袈裟的衣纹质感刻画上精细入微，在直线与波浪线的对比下仿佛都能感受到透明度，生动至极，在雕塑空间节奏控制上，果断、概括、犀利，特别是背景墙与形体接触边线的强弱、远近、高低变化处理之美，难于言表，就是在享受雕塑艺术空间。虽然头和手遭到了不同程度的破坏，但它不影响整体雕塑的欣赏价值。这种浮雕形式具有草原民族性格和欧洲文明碰撞出的艺术风格雕塑形式，也是古丝绸之路文明促进北方草原文明繁荣发展的历史见证，凸显了辽代艺术家们对艺术本体美的追求和进取精神，为辽代的雕塑艺术注入了新鲜的血液。

还有在内蒙古呼和浩特华严经塔上的天王，最富有表现力。虽然残缺，但是他形体的生动让你看了忘不掉，它就是艺术力量之美，并且有古罗马、意大利文艺复兴时期雕塑艺术一样的魅力，仿佛更鲜活、生动。雕塑张力能触及你的灵魂（插图12）。显然雕塑思想和审美受到西方古罗马时期雕塑世俗化精神的影响，讲究科学，有人一样完美的比例，健美的形体、夸张的肌肉刻画出人一样的温度，看上去就像是世界级的健美运动员雕像，在中国雕塑史上从来没有呈现过像这样震撼的雕塑艺术。它突破了传统宗教里被神化了、程式化了的天王。传统天王在宗教里象征着对邪恶震慑的力量，而辽代创造出人一样的天王不仅象征对邪恶震慑，还有象征北方草原民族英雄保护大辽平安的意义。

从塔上的天王雕塑看，辽代告别了传统被神化的天王，创造出代表北方草原民族精神的英雄形象。健美的肌肉抖动，仿佛能听到咔咔的声音。运动形态把握如此之精准，从胯到腰、从腿到脚的力度顶天立地。这条形态的曲线太完美了，还有腰胯结构表现使形体动态达到了出神的境界。塑造的形体大气、豪放、雄浑，有强烈的视觉冲击力。尤其是在塔的东面有一个塔梯可以上到塔身一层，能更近距离欣赏到两侧的天王身躯。那一刻都不敢相信，这是在欣赏中国古代雕塑吗？完全被雕塑的魅力所征服，好像是在欣赏欧洲古罗马和文艺复兴时期大师们的雕塑作品（插图13），瞬间天王的概念已不存在，只有雕塑艺术的存在，那厚重的塑造、生动的塑痕、概括的形体、流动的线条，从侧面看被拉深

插图12
天王
塔身一层砖雕
内蒙古呼和浩特华严经塔

插图13
奥古斯都·屋大维
大理石雕刻
古罗马

的空间，富有弹性的肌肉，使形体更加立体生动。从塔一层北面的天王，除头和手被破坏掉，余均保存，是这座塔上比较完整的一件天王雕塑，形体张力表现非常生动，艺术家们的激情不加掩饰地全部在雕塑作品里释放出来，气势磅礴，飘带的设计很巧妙地在形体和空间中穿梭，游刃有余地将刚柔对比关系烘托得栩栩如生，使天王形态结构本质的美更加突出。由装饰性飘带、短裤、腰带、披肩，线性的结构与形体节奏上的肌肉力量融合在一起产生玄妙的和声，构成一部完美的"力量交响曲"，这是对力量的赞美，也是对人类智慧的表扬，这又是一种雕塑形式的诞生。艺术家们用写实的手法、浪漫的思想、诗性的激情，创造出具有草原民族性格独特的艺术风格雕塑，彰显了辽代艺术家们的自信、敢于创造的精神和艺术审美。

如果说华严经塔上的天王是对力量的赞美，那么菩萨就是对形体美力量的歌颂。

从塔身一层西南面的菩萨看，虽然破坏很严重，头和手没有了，整个形体还算比较完整地被保留了下来。但菩萨的雕塑艺术魅力不减，它也有着天王一样的力量，就是形体美的力量。不再像传统的菩萨程式化、呆板的造型，而是新的创造，有人一样形体的比例和温度。雕塑的生动、饱满、大气、自然、舒展、流畅，特别是在形体塑造上很注意生命形态的刻画，从腰与胯结构的穿插，左腿微微往里扣，重心点在左脚上，右腿放松构成了形体动态曲线（插图14）。完美的曲线使整个形体上的特征完全绽放出来，这条富有弹性的力，也是人们常说的"维纳斯"之"美"，她在希腊的名字叫阿弗洛狄德，罗马的名字叫维纳斯（插图15），是美和爱之女神的象征，其形体比例九个头，表达的是"神"之"美"。而辽代艺术家们创造的菩萨形体比例是八个头，科学地将人之美表达出来。无论是"神"还是"人"之美，都是人类共同追求美的理想创造，也是人类对美的形态的赞扬。辽代的艺术家们准确地捕捉到了生命运动规律，使运动曲线凸显出艺术美的力量。在今天的现实生活中继续上演这种姿态，如：世界时尚T台秀上的模特们展示个人形体风采亮

相的姿态，就是辽代艺术家们发现这种姿态将它创造出来，成为艺术"美"的永恒力量。

内蒙古宁城（辽中京）大明塔的雕塑艺术独具特色，是绘画与雕塑相结合的艺术形式，浮雕加敷色彩。这种形式继承了石窟中的彩绘石刻形式，但它有所不同，由于色彩与砖灰色混搭在一起，浮雕的语言被淡化，绘画的效果被凸显。从塔身上西北面的天王看，形体动态更注重神似，夸张肌肉的力量，注意面部每一个细节的刻画，从表情肌到瞪大双眼的表情结构，从张开大嘴到嘴的表情结构的深入刻画。这种入微的表现，是让邪恶看清，如果是犯下罪恶将会粉身碎骨，被绳之以法。还有的天王面部表现紧锁眉头，双眼注视前方，双唇紧闭表达对邪恶的威严力量。在雕塑形体上的佩饰，项圈、璎珞、披肩、天衣、裳裙、带饰刻画得精细入微，很像中国画里的白描，再加上雕塑材料的灰砖本色和褪了色的土红、石青、石绿，夹杂着保护层的白色混合在一起产生一种现代中国画的人物工笔重彩、外带点泼彩效果，很惊艳地将天王的神采表达得出神入化。这种独特的视觉效果，如果运用到中国画人物创作里，将会产生意想不到的震撼。辽塔雕塑艺术释放出美学思想的魅力仿佛是取之不尽的源泉，看后总会受到启发。这是辽塔雕塑艺术与其他朝代艺术不同的地方，它充满活力。

辽砖塔独特的雕塑艺术演进了辽代雕塑艺术发展史的全过程，像一个永不停息的生命，一直在生长，这就是辽代草原民族留给我们今天的艺术力量。

一路上搜集的辽代砖塔雕塑艺术带着时间、岁月、历史的沧桑。还有珍贵的艺术文化历史遗产，从传统到现代，从继承到发展，从宗教到世俗，从程式化到个性化，从神到人，从吸纳到创造，从平面到立体，从装饰到写真，从室内到室外，创造出中国最早的现实主义雕塑艺术，形式多样，技艺精湛，雕塑语言内容丰富，有创造性的思想和技术，独特的草原民族文化艺术极大地丰富了中国文化艺术发展史的宝库。

还有一项不可忽略的残缺雕塑艺术，它以超越历史本身的姿态向我们这个时代走来。它的艺术审美更富有生命力，里面有辽代艺术家们伟大创造的文化积淀与历史、时间融为一体形成的新艺术语言形式，这是辽砖塔雕塑艺术留给我们的又一种文化遗产。这种新语言更能启发艺术创造的想象力和激情，当代艺术家们已经把这种残缺的美学思想和艺术形式直接运用在艺术创作里，产生了奇妙的艺术视觉效果和思想。这是完整形象无法代替的艺术审美。

辽代砖塔雕塑艺术证明只有吸纳接受先进的思想，敢于突破创造，艺术才能繁荣昌盛，创造出更具有民族风格、历史价值的艺术。

这所有的收获和挖掘都是由徐秉琨先生整理与归纳包恩梨先生的三十年前精准详细的记录，在徐秉琨先生一路上的指导下，才有了今天这样全面丰硕的收获，这也体现了徐秉琨先生、包恩梨先生对北方草原文明的热爱和对历史的责任、奉献精神。使我们今天才能看到在北方草原上，曾有过一段辉煌草原民族文明发展的历史。如果没有徐秉琨先生的坚持让我来完成这次的任务，我将永远都不会触碰到这段伟大的历史，也不可能知道辽代的雕塑艺术如此的丰富和辉煌。我们的挖掘和发现也是让大家了解在中华民族发展史上曾经有过一段草原文明发展的光荣历史。是草原民族创造的马镫文化，使欧洲提前进入了现代社会，而欧洲文明也促进了草原文明历史的发展。砖雕塑艺术里面就有草原文明与欧洲文明相互学习的历史见证。因此它不仅是中国草原文明发展史的物证，也是世界文明发展史的物证。它当之无愧属于世界文化遗产。我们每个人都有责任不让祖国的瑰宝在我们这个时代再没有了。希望国家能采取一些保护措施，不要让草原文明变成历史的遗憾。感谢徐秉琨先生、包恩梨先生给了我这次机会，我也不知道我的工作能不能使他们满意，但我已经尽全力了，也感谢一路上和我一起搜集资料的好友张守国先生甘愿牺牲个人时间的奉献精神，还有一路上热心帮助的朋友，如果没有他们，也不可能搜集到这么全面的砖塔雕塑艺术珍贵的历史资料。能为中国雕塑艺术发展史找回被遗忘的一段历史，我们所付出的一切值了。

郑波
2017年4月29日于沈阳

SEARCHING FOR THE LIAO CIVILIZATION AND DISCOVERING THE NEGLECTED BRICK-PAGODA SCULPTURE ART

In order to supplement some pictures for Prof. Bao Enli's manuscript—*Sculpture Art of the Liao Dynasty*, Professor Xu Bingkun consulted the notes that Prof. Bao had taken thirty years ago while making a field trip to inspect the Liao Civilization, compiled a list of entries based on the manuscript, and commissioned me to take photographs accordingly.

I started the preparation work including choosing candidates for photography, camera devices, transportation vehicles, itinerary, navigators, and a compass. The compass is a very important tool in this trip. Besides its application to the itinerary, the compass also has an important function of marking pagoda's orientation at the shooting scene, which is the most basic requirement of archaeological work. Therefore, this is a goal-oriented and highly professional task.

First of all, we should search for these cultural relics in the territory of Liao Dynasty, involving areas like the Upper Capital of Liao Dynasty (Balin Zuoqi, Inner Mongolia now), the Middle Capital of Liao Dynasty (Ningcheng, Inner Mongolia now), the Western Capital of Liao Dynasty (Datong, Shanxi now), the Southern Capital of Liao Dynasty (suburb of Beijing now), the Eastern Capital of Liao Dynasty (Liaoyang, Liaoning now), and also including museums of cities, prefectures, and counties, temples, pagodas of Liao Dynasty, cave temples, grottoes, and etc..

Meanwhile, it is a meticulous archaeological survey work. Taking one of the entries as an exemple, its description of objects has corresponding requirements for photography work, 'Name: eight—sided Buddhist stone pillar, sumeru podium, and the hole on the rock under the lotus seat. Comment: the pillar is engraved with Dharani scriptures on seven sides and names of the engravers such as monk Yunchong, as well as the date of construction are on one side. Each side of the sumeru podium is engraved with dragon, phoenix, peony, lion, kalavinka, and so on. On the rock under the lotus seat, there are six holes that are 130 centimeters apart from each other arraying with hexagon shape, 7 centimeters in diameter and 2 centimeters in depth for each hole. They are probably plinth cavities left from the hexagonal pavilion at that time.'

I was a fine-artist but a layman in archaeology, so I felt puzzled and worried when Professor Xu Bingkun chose me to complete this arduous task. This was undoubtedly a huge challenge and I could only go with curiosity.

Since I accepted the task, I should take responsibility. First of all, I needed a companion with professional skills, as well as spirit of voluntary and devotion. The very first person I thought of was my good friend Professor Zhang Shouguo, who was quite professional and ambitious. He had a twenty years' experience of photographing in Xizang, and at the same time, was a member of the China Photographer Association. When I contacted him and told him about the task, he accepted it immediately and was ready to set up and drive there at any giving moment. After all these preparations, we started the journey.

In July 2016, we began to travel through time and space, entering the Liao Dynasty.

We encountered a lot of problems at the beginning of our trip. There were too many things we didn't think about: the navigation was difficult to use and we had no idea where to approach the mountain. Once we began our journey, the bitterness began. Sometimes, we had to push the car instead of the car carrying us. There were only two of us so we had to rely on each other. We were exhausted before reaching our destination. Sometimes, the road ended at a trail set up by hillmen for gathering mushroom, and when we drove through it and reached the end, we were wet with cold sweat. A little further away was the valley. We had to walk on the ancient plank road in order to get close to the subject at places like this. What's more, we had to take equipment with us, which weighed more than twenty kilograms. Relics like temples, grottoes, and some pagodas were all photographed in such environment, so this was a dangerous work. The pagodas were the most difficult to shoot, especially those in the mountains. We had to take risks to photograph the contents of the tower since the shooting space was quite limited and the base was two meters wide at most. Most pagodas were built on top of the mountain or on the mountainside, so we had to go back down to shoot the whole scene, which often challenged our limits. Usually, a person behind held a person at the front to complete the shooting. Tumbling and falling happened all the time. Sometimes, we hurt our arms, hands, and legs during the work, but when we returned to our accommodation, we even did not dare to take off our clothes because the blood had already stuck to them and it took us a few days to recover. We have all experienced car tire bursts, and car repairs, and we were quite stressed out for the busy schedule. This journey became an adventure.

The moment when we captured the Liao Dynasty civilization, all the bitterness and hardships vanished into smoke. We wanted to shoot more and became excited like warriors attending a battle. Though the goal was clear, sometimes we had to act quickly or wait for a few days to start the work since the shooting was largely related to time, weather, light, and angle. The ultimate goal, of course, was to collect the relics of the Liao Dynasty and to complete the task that Professor Xu and Professor Bao have given us.

Along the way, we collected materials about sculptures of more than twenty brick pagodas and found the neglected sculpture art of Liao Dynasty. As a professional art worker, and from the history of art, I only know the Qin and Han Dynasties, the Wei, Jin, Northern and Southern Dynasties, the Sui and Tang Dynasties, the Five Dynasties, the Song, Yuan, Ming, Qing Dynasties, but have no knowledge of the Liao. From the ancient Buddhist sculpture art, there are well-known

Mogao Caves in Dunhuang, Maijishan Grottoes, Longmen Grottoes, Yungang Grottoes, and etc.. The Chinese cultural scholars must have contributed a lot in this area, for even those who do not study art know these grottoes. However, they pay little attention to the culture of the northern minorities. The Liao Dynasty also has a glorious history of more than 200 years. In addition, the grassland ethnics once has influenced the world. According to western world historians, the stirrups created by China's grassland ethnics have brought Europe into the modern society ahead of time. This history should not be neglected.

Shooting the ruins of Zuzhou City in the Upper Capital of Liao Dynasty (Balin Zuoqi now) was the windfall of this trip. After we photographed the collections of Balin Zuoqi Musuem, Professor Wang Weixiang, the curator, brought us to Zuzhou City of Liao Dynasty to visit the 'Stone House' built in the 4th century for there was still some time left. As I was climbing up the mountain, I saw the 'Stone House', and was immediately sturned by the sight (Fig.1, Fig.2). What I saw was a modern sculpture with perfect surroundings. How amazing it was! Just as the sculptor Constantin Brancusi (the most influential sculptor in the 20th century, known as the pioneer of modernistic sculpture) says, 'architecture is an inhabitec sculpture'. These words were reinforced at that moment.

Fig.1
Stone House
front side and the totem
built in the 4th century

Fig.2
Stone House
'chamber-angle structure' of the building
built in the 4th century
Zu Zhou City of Liao Dynasty, Balin Zuoqi

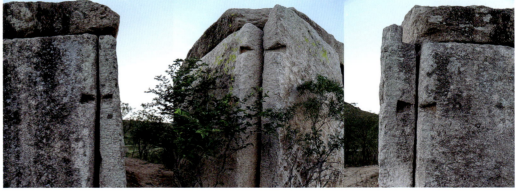

Looking out from the house, I saw a spectacular and poetic picture, with vast grassland and rolling mountains shining in the sunset glow. Against the background of magnificent mountains, 'the Stone House' seems to have thoughts and personalities. This seven-huge-stone-constructed architecture itself is a miracle.

If observed carefully, every piece of stone has its structure. The Stone House is built with high technics to dissolve the prairie hurricane and fight against natural disasters, as its groove and its slant are perfectly filled together, unchanged. The unique door of the Stone House, a grand opening and closing atmosphere, has shaped the rough and firm character of the Khitan people. Inside lies a 4.3-meter-long and 2.5-meter-wide stone slab flat on the ground, which the museum experts refer to it as a bed. The slab is also found to have nine-square grid on both its east and west side. Two engraved lines go from east to west pointing to the middle, where it has symbols of different shapes. The patterns on the west seem like ancient divination figures, which are related to Astronomy, play an important role in grassland travelling and patriarch discussion. In short, this makes the 'Stone House' more mysterious. The gap between the northern wall and the roof, together with the southern door, effectively improves ventilation. Both the outer wall and the south roof the building are engraved with patterns or characters. Though they are not clear, the structure of the graph can still be recognized. There are also human-shaped totems on the walls. I told Professor Wang what I saw, but he didn't see them before and said it was the first time that he had heard someone say there was a totem on it.

'It's impossible,' he said. 'I have worked here for decades and been here for countless times, but I've never heard of totem.' I pointed out the totem to him. He was excited and said, 'This is a new discovery.' I told him how I had discovered it. Unlike ordinary buildings, the house is uniquely constructed. It's not simply a 'stone room' or a 'stone house' just as archaeologists call, but it is a 'hieron' or 'shrine' more accurately. So I searched every detail of the building with doubt and curiosity, and finally found the totem.

In order to obtain this information, first of all, particular conditions should be met, such as specific time, light, and weather. The most important thing is that the discoverer should bear them in mind. This is what I've learned after inspecting so many ancient rock paintings. It happened to meet all the conditions that day, that's why we were able to see it. I wonder if it's luck or our efforts have moved the God to let us know so many stories. Perhaps those totems are eager to communicate with modern humans and wish to convey their stories to more people. The 'Stone House' is a masterpiece created by Khitan and has high academic and practical application value for studying ancient culture, architecture, mechanics, environmental science, folklore, aesthetics, and the modern art in particular.

In the Balin Zuoqi Museum, we captured a lot of sculptures of northern grassland ethnic group. We were shocked by one of the Khitan character stone carvings (Fig.3). It is a round sculpture work, and as tall as a person. Though it lost its head and one arm, it is still brilliant. The sculpture is vigorous and the figure of the character is precise, smooth, vivid, and brief but not lacking details. It slightly portrays the features of the costume to fully exploit the stone material, with the remaining hands folded. In the eyes of archaeologists, this sculpture is defective. However, in the eyes of the artists, it is perfect. The imperfection of the sculpture has generated a mysterious artistic effect to endow imagination and inspire the recreation of art. This creation reflects Khitan artists' comprehension on stone material not simply imitating a person, but creating the eternal art with

Fig.3
Weng Zhong: stone statue placed in the tomb passage of
the Liao Zuling Mausoleum, Balin Zuoqi
clearing in 2003

a living stone. I think Henry Moore, the master of the modern sculpture, would be touched by it, since it represents both modernism and primitive spirit. The artists of Liao Dynasty inherit and carry forward the spirit of Khitan people. Sculptures of the Liao, especially carving on brick pagodas vividly showed the development history of the Liao sculpture art, and created a three-dimensional and lively ideology development history of Liao Dynasty's Culture and Art.

The Jin Pagoda in Simu, Haicheng, Liaoning Province, is the focus of our photography work. We found the Jin Pagoda according to the clue given by the entry. The entry wrote, 'The No. 38 Jin Pagoda of Liao Dynasty has a niche on each side, with a Buddha in each niche and two Bodhisattvas outside the niche on the left and right. The Bodhisattva is made of bricks of 10–20 layers with its body engraved in semi-relief, while its head, constrcted of 4 layers of bricks, sculptured in the round and poke of the main pagoda. The No. 40 Jin Pagoda has a Bodhisattva on the north side of the tower. On top of both Buddha and Bodhisattva assemble brick carving canopies, and on each canopy, there is a pair of flying Apsaras... They are holding things to offer, dancing, and playing music... No. 35 Jin Pagoda has a total of 48 figures of brick carving including musicians, dancers, and messengers around niched doors on the west side of sumeru podium's Ying–xiang pillars. Some binding up their trousers, wearing necklaces of jade and pearls, are holding things to offer; Some are sitting on the knee and fixing the guzheng; some are playing mongol stringed instrument; some are holding Buddhist scripture, sticks, or feathers; some are holding or carrying wine jars; some are playing the pipa or the flute; some are dancing wearing Khitan clothes; some called anupadaka (self-created) children are shouldering the word '人 (human)' or '我 (self)'...' This is part of the descriptions of the pagodas summarized by Professor Xu from the field study notes written by Professor Bao thirty years ago.

When we approached the Jin Pagoda, we first saw a magnificent and vigorous architecture with European Roman style. The octahedral cylindrical building stood out in the green mountains under the sun, looking just like its name—golden(Jin) pagoda. How beautiful it is! When we came closer, we were shocked! Unlike what has been described in the notes, only the flying Apsaras and the guardian lions are preserved, and all the other things are destroyed, without even a single Buddha left. All the Bodhisattvas lost their heads and part of their arms. None of 48 statues of brick carvings are complete. I reported the situation to Prof. Xu and sent photos to him. Prof. Xu sighed deeply, '...It's too late...' Then I sent photos of those guardian lions to Prof. Xu, and he sounded very excited. 'Great! Great!' he said. 'This definitely has never been seen in China before. The expression of intramuscular block is like a Roman sculpture, it's too tense.'

Professor Xu urged me to shoot carefully, and then I sent some photos of flying Apsaras to him. He again said to me with excitement, 'I have never seen such beautiful flying Apsaras before. They look like the Tang Dynasty style.' Likewise, he told me to take more photos and pay more attention to the production technique in these sculptures. The following work was easy to carry out. We could photograph what had artistic value freely instead of searching according to the entry. The most difficult scene to shoot was the pair of flying Apsaras on the north side of pagoda, which was fairly high. Below was the valley, and we had no space to step back, so we took a bold action in order to photograph the whole sculpture. Firstly, I returned to the mountainside and grabbed the tree branches to see whether there was still some room to expand. I explored everything in advance and let Prof. Zhang set the tripod up in front of me. I protected him with my body and he photographed, and eventually we made it. This pair of Asparas is the most beautiful one. From its modeling, we clearly see that it has a Tang Dynasty legacy and is significantly influenced by the Tang. Their physiques are plump, free, and aesthetic, and heads look much like those of three-color glazed pottery figure of noblewomen. The expression of the pair is general, bold, and lively with realistic technique. This pair of flying Apsaras, sculptured in the half round, is excellent work among Chinese Buddhist flying Apsaras.

There are still several guardian lions preserved well on the Jin Pagoda sumeru podium. They are quite shocking—realistic, bold, plump, powerful, and present exaggerated muscles and physical strength. The decorative flowing lines make their hair look vivid. The smiles on their faces look humorous, which seem to tell vistors that it isn't a big deal to lift the pagoda. The action designs of these guardian lions differ from one another. Some lean their heads out of the sumeru podium, looking agile, some lift the pagoda with arms or shoulders. The facial expression is relaxing, but the action design is powerful. The muscle symbolizes power and the smile symbolizes kindness. On the whole, the guardian lions are friendly. They are an expression of the true, the good, and the beautiful, and they also represent power against evil. These seem to be more secular and described in modern language more down to earth. The artists of Liao Dynasty interpreted religion with realistic technique and romantic thoughts, and created the unique civilization of the grassland ethnics of Liao Dynasty. The strong men are used the same technique as the guardian lions. For the action design, some lift the pagoda with shoulder, some with neck, and some with hand, all with a vivid smile on their faces. This shows the artists' enthusiasm and their high artistic appreciation of the sculpture.

From the perspective of a damaged head from the strong man, it portrays a subtle, concise, unrestrained, and vivid image, and lively shaping the human spirit (Fig.4). The muddy taste of the head texture reminds me of Balzac's head sculptured by famous French sculptor Rodin (Fig.5). They are so much alike as if they were created by the same sculptor, although one was created between 10th to 12th century in Asia, and the other was created during 19th century in Europe.

Fig.4
Strong Man
brick carving in sumeru podium
Jin Pagoda in Simu, Haicheng, Liaoning Province

Fig.5
Balzac(part)
Bronze(1893–1897)

The hands of those imperfect Bodhisattvas vividly show skillful techniques of artists, as well as their passion to create. It seems that we could appreciate the whole artist's process of creating sculptures, with the action of adding mud, slapping, and molding, vividly interpreting the taste of mud to the fullest. These damaged works left from the pagoda fully show superb techniques and artistic aesthetics of the artists of Liao Dynasty. They differ a lot from the Buddhist sculptures of the previous dynasties in materials, places, spaces, and aesthetics. Under normal circumstances, sculptors usually used the mixture of clay and water to create the bozzetto. They shaped the image of what they wanted first, and then chose materials for the creation, such as bronze, stone, wood, and so on.

Mud draft is a testing process for artists. It can be created and modified freely according to artists' thoughts. When it fully achieves the optimal results of what artists want, the draft is completed. Then artists convert mud draft with materials they need, whether to magnify or contract, and whether to use stone, wood, or bronze. Therefore, the original mud draft is used in the early stage of sculpture making. However, the artists of Liao Dynasty directly burn the mud draft in the brick kiln and lay it onto the pagoda, and then such a piece of work is completed. It is the reduction of the replicating process that preserves the original enthusiasm of the sculpture. Through those damaged works which their protective layers have been lost, we can still see the sculpture techniques, such as slapping, engraving, sculpturing, digging, troweling, and rubbing into different shapes of strips, ropes, balls, and so on. They are directly used in sculpture and freely used by artists to easily release their personal thoughts.

These techniques play an important role in creating the unique style of sculpture art in the Liao Dynasty. The most important feature of the Jin Pagoda is embodied in the character 'sculpting', which highlights aesthetic taste and individuality of the sculptors in the Liao Dynasty. Another material that can not be ignored in the pagoda sculpturing is the adhesive used for laying bricks. The adhesive is made of a mixture of lime powder and glutinous rice flour, which is equivalent to the cement invented by the ancient Romans. The grassland ethnics of Liao Dynasty grind the lime powder and the glutinous rice flour more carefully, blend them with different proportion, and apply them to pagodas. The surface of the sculpture should also be applied. It is not just a simple application, but a second creation carried out by artists, using it to 'find' the damages during firing and transportion. At this moment, the sculpture mark left by artists during the creation process plays an important role, which can not only let the adhesive cling to it, but also modifies and polishes the sculpture to make the work more solid, vivid and perfect. This kind of material has experienced thousands of years of testing, but still remains perfect (Fig.6).

Meanwhile, we found that the Liao pagodas had different origins of materials due to their different geographical locations. Though they are all mixture of lime powder and glutinous rice flour, the surface textures created by them are different. Some look like jade carving, some glazed pottery, and some marble carving. Such rich surface effects can all be seen in the sculpture of the brick pagoda in the Liao Dynasty. The technique created by the grassland ethnics of Liao Dynasty is still used in Italy nowadays. The world famous Italian sculptor and artist, Bruno Walpoth, regarded as one of the greatest woodcarving artists of the century, applies this technique in his woodcarvings (Fig.7, Fig.8). He applied the adhesive to the surface during the process of sculpting and after finishing the work. The repeated application during the creation process was a process of modifying and polishing, which not only kept the traces of the sculpturing process, but also smoothed the visual effect of the works. This technique, which formed Bruno's unique personal style, was widely used in the sculpture art of the Liao brick pagoda.

The sculpture art of the Jin Pagoda reflects the wisdom of the Khitan people and develops the

Fig.6
The King of Heaven
brick carving
Pagoda of Jingyan Temple in Kazuo,
Liaoning

Fig.7
Girl (1)
contemporary wood carving
Bruno Walpoth (Italy)

Fig.8
Girl (2)
contemporary wood carving
Bruno Walpoth (Italy)

Buddhist Sculpture Art. The clay sculptures and stone carvings of the temples and the grottoes that we usually see before the Liao Dynasty, all belong to indoor sculpture. The sculptures of the brick pagoda in the Liao Dynasty were made outdoors, with different materials, location, spaces, and artistic aesthetics compared to the previous dynasties. These sculptures highlight the vivid, unrestrained and casual characteristics of the grassland ethnics. When Buddhist sculpture developed from indoor sculpture to the outdoor, it was also the time for the development of sculpture art and the mental emancipation of artists in the Liao Dynasty. The brick pagoda sculpture reflects the northern grassland people's characteristics of freedom, boldness, confidence, bravery, imagination, creativity, and passion towards life. It has created brilliance of outdoor sculpture art in ancient China.

When collecting sculptures of more than twenty Liao brick pagodas that were found together, we could see the unprecedented development of the decorative sculpture in the Liao Dynasty. From a theme relief— 'flying Asparas', their artistic expression are rich in form and have extremely high artistic level. There are three different kinds of semi-round sculpture, high relief, and bas-relief, with techniques including realism, decoration, use of lines or colors. This is not only the complete collection of decorative relief art in the Liao Dynasty, but also the most kinds of decorative sculpture in Chinese history. When I was excitedly appreciating the 'flying Apsaras', I seemed to have seen artists of Liao Dynasty discuss beauty of art and law of motion. For instance, the composition and the shape of a pair of 'flying Apsaras' carved on the brick pagoda in Guangji temple of Jinzhou city differ from that of the traditional. The composition of the flying Apsaras is triangular in shape, which is surrounded by a great cloud-knitted arc. The arc generates such a huge sense of vibration that the flying Apsaras seem to have a force to glide forward, forming a majestic composition. The image of the flying Apsaras seems to lack the Buddha nature of the past, but more closely resembles people in reality. They look more friendly and peaceful, with modern-sensed hair styles, and are shaped like humans with warmth, vividness, summarization, as if they have been given new life.

In terms of sculptural Language, the shaping and compression between the chest, hips, and knees of flying Asparas, as well as the control of high and low rhythms, are extremely perfect, And these elements make the forming of flying Asparas just right in grandeur, thickness and quantitative change. The artists skillfully combined the flying Asparas and the clouds with decorative lines, which does not damage their body, but instead, strengthens the freeness of their gliding posture. The artists pay great attention to details of the image of flying Asparas. For example, the posture of two legs, the upper leg pressing upon the lower, and the lower leg's foot on the upper one, produces a body twist, giving a dynamic change to the triangle and emphasizing the elegance and stretched physiques of the flying Asparas. In particular, the changes of lines on shawls, skirts, necklaces of jade and pearls, clothes, and belts are perfectly sculptured, which shows the artists' perfect control of width, density, depth, level, arc, and straight line. Thus, the language of lines is the soul of the sculpture language of the flying Asparas. It shows the creativity of artists in the Liao Dynasty and brings people's thoughts into religion. This pair of gliding 'flying Asparas' is a good proof that the inspiration for artists should come from the life on the northern grasslands. They vividly convey powerfulness, elegance, and free spirit of eagles soaring on the grassland. The creation of the gliding 'flying Asparas' symbolizes the aggressiveness, bravery, and pursuit of beauty and freedom of the grassland ethnics, and at the same time demonstrates their wisdom and imagination. Now Let's enjoy the flying Asparas, which symbolizes the spirt of the northern grassland ethnic of the Liao. Although human beings have realized the dream of flying in the sky, just like the winged flight of extreme sports in the world, they have never stopped the pursuit of beauty and freedom. The arts and ideas created by artists of the Liao Dynasty will continue to evolve in our time.

The spirit and aesthetics of the grassland ethnics are integrated into not only the 'flying Asparas', but also every sculpture of the brick pagoda in the Guangji Temple, Jinzhou. For example, the Bodhisattvas on the pagoda were more typical and each Bodhisattva had its own characteristics. The pair of Bodhisattvas in the west looked bold, unrestrained, and vivid, as if they were created to be national heroes. Unlike most traditional Bodhisattva, they looked handsome, kind, and wise, and giving people a sense of intimacy. You could not even tell that they are Bodhisattvas, and only the kasaya could help to identify their identities. From the perspective of sculpture, the traditional norm and aesthetic sense of tranquility have disappeared, and a new image has been created. Especially in the sculpture language. the strong carving technique highlighted artists' confidence and artistic ideas. The crazy carved lines and the skillful carving technique had made the brick sculpture art reach to an extremely high level.

From the Bodhisattva's head, the appliation of carving techniques, including pressing, tapping, carving and cutting, is very effortless. The knife technique is clean and sharp, and the image is firmly shaped. And its texture is similar to stone carving, summarizing, vivid and full, transforming the muddy taste of brick sculpture into the world of carving techniques. The carving technique used on Bodhisattva's kasaya was more wild, free, and full expression of linearity in sculpture, and even a

Fig.9
The bust of Diego
bronze sculpture(1950)
Alberto Giacometti(Switzerland)

Fig.10
Bass-relief
bronze(1932)
Marino Marini(Italy)

straight line could not be found. At this moment, artists' enthusiasm was released and people could imagine a picture of artists humming a prairie tune, shaking shoulders, and dancing the unique victory horse-step-dance of the khitans to pay tribute to heroes. What a fascinating line carving!

Obviously, the vivid line carving derives from artists' observation of reality, which is no longer a traditional pattern. Though straight line cannot be found in the sculpture, yet artists create a real and vivid image, and those elegant curves express objective existence. We could imagine that artists were shaping an image of a real hero standing high in the breeze. The visual effect of his clothes flapping his body reflects artists' observation on fine details, especially those with regards to the depiction of life forms. For example, the knees of two legs slightly buckled inward perfectly captures the key point to vividly manifest the firmness from head to toe. In addition, the whole body looks tall and straight accompanied by cassock and vivid carving line. This is not only a new image created by artists using realistic techniques, but more importantly, it pushes the traditional Chinese religious sculpture art forward into a new era. It is a watershed of traditional sculpture and modern sculpture, which symbolizes the real entry into the phase of sculpture creation of realism.

Fig.11
Bodhisattva
brick carving
Pagoda of Guangji Temple, Jinzhou, Liaoning Province

The brick pagoda sculptures of Liao Dynasty give rise to the integration of different national civilizations, accompanied by the emergence of various sculpture styles. Artists are brave enough to stress on the art noumenon and explore the artistic aesthetics through sculpture techniques, creating unique sculpture artworks that have never seen before. Actually, the challenges posed by such knife skills on the traditions had been unparalleled, but it is a wonderful art form favored by modern sculptors. Alberto Giacometti, the existentialism sculptor of Switzerland, and Marino Marini, the surrealism sculptor of Italy, were both masters of the 20th Century who interpreted the art through knife skills, offering unique insights that greatly contributed to the development of sculptures around the world (Fig.9, Fig.10). However, the sculptors of the Liao Dynasty successfully demonstrated the aesthetics of art through knife skills as early as the 10th to 12th Century (Fig.11). What's greater is that the artists created the very first existentialism sculpture of China with knife skills, creating a new era in the history of Chinese sculpture. There is no doubt that such achievement is tightly connected to the great age in which the artists lived, as the historical glories were demonstrated by the brick tower sculptures of the Liao Dynasty.

The Liao pagoda sculpture art has explained the thought of religion with its unique national spirit, and also created the culture of the grassland ethnics. The sculpture art on the Liao pagoda, collected along the way, is permeated with traces of grassland culture everywhere. There appears the image of the 'horse' under the lotus seats of the Buddha in the tower such as the North Pagoda in Chaoyang, the Yunjiesi Square Pagoda on Fenghuang Mountain, and the Wangtuzigou Great Pagoda in Chaoyang, and so on. The image of the 'horse' also appears in the sumeru podium of Pagoda of the Kazuo Jingyan Temple, Chaoyang, Liaoning Province and the story of the practitioners. The 'horse' represents the spirit of the northern grassland ethnics, and it symbolizes the love and worship of horses by those groups. It is also the most precious gift for the grassland people.

There are many people or things on the Liao pagoda that reflect the characteristic of the grassland ethnic group, such as the wearing, leather and clothing worn by figures on the sumeru podium of the Yunjusi North Pagoda on Shiing Hill, Fangshan District, Beijing. These are the characteristics of the real life of the northern grassland ethnics recorded by the artists with their exquisite craftsmanship and realistic technique. It is also the creation of the artists for religious culture.

From the sculpture art of the Liao pagoda, we can also understand the open mind of grassland ethnics, who focus on learning, absorbing, communicating and inheriting the advanced ideas and technologies of different nationalities.

Many sculptures on pagodas have Tang Dynasty characteristics, for example, the sculpture style and content of the Jin Pagoda of Haicheng, Liaoning province have features of the Tang, and on the sumeru podium there are also emissaries wearing Tang's official clothes. The Balengguan Pagoda in Chaoyang is another manifestation of Tang Dynasty culture which reflects the worship and awe from people in the Liao Dynasty towards the Tang culture. Distinct from pagodas previously introduced, both the Buddhist niches and statues are exposed out of the pagoda, where lotus seats are designed in a large semicircular structure and the legs of Buddhist statues are sculpture in the round, endowing the statues with the style of Tang Dynasty. Exquisite designs, which can be found on a remnant Buddhist statue, were applied to various locations ranging from its garments to the lotus seats. The kasaya and clothes on the statue are designed to well fit the variations of body parts. With the draperies all missed and only the physical shape and regular patterns left, the overall image of the statue give impressions of vigorousness, richness, conceptualization and aestheticism, making it a typical decorative statue. This is also a rare piece of sculptural work with Tang Dynasty artistic features and of exceptionally high quality and artistry. Unfortunately, the severe damages imposed on the statue have left both its head and hands missing, and such a great artistic sculpture can only be pieced together in our imagination. In the meantime, the content provided in the sculpture on sumeru podium is like a picture-story book reflecting the royal court life of Tang Dynasty, in which attendants, musicians and dancers are dressed in clothing of Tang style, while the structural compositions and images on the engraving of musical instruments and tools are realistically lifelike. Unfortunately, all of these have been severely damaged, to an extent that only certain shapes and patterns can be recognized from the remaining marks. Otherwise, we may still be able to get a glimpse of the life of the great Tang period in the eyes of people living in the Liao Dynasty, and even more so, appreciate the brilliance of sculptural art in the Liao Dynasty.

The sculptures include figures of Buddhas, Bodhisattvas, flying Apsaras, and priests of the Taoism, collected at the Balin Zuoqi Museum of Inner Mongolia, all of which are original works from the South Pagoda of Liao's Upper Capital. The fact that so many cultures and thoughts have been integrated in a Buddhist pagoda proves the cultural prosperity in the Upper Capital of Liao at the time. The winged flying Asparas, as well as envoys of different ethnics, providers and western emissaries were also discovered in the Qingzhou White Pagoda of Balin Zuoqi. At the sumeru podium of the Square Pagoda in the Yunjie Temple, located on Fenghuang Mountain in Chaoyang, there were also images of western envoys with wavy hair and beards. Similarly, flying Apsaras, envoys from Europe and Asia, and musicians, etc., can also be found at the sumeru podium of the North Pagoda in the Yunju Temple on Shijing Hill, Fangshan District, Beijing. These examples have indicated the openness, inclusiveness and progressiveness of the cultural exchanges occurred in the Liao Dynasty. Yet all these would not have been possible without the most intuitive and realistic records accomplished by the artists using sculptural techniques, which are still of great historical and realistic significance for today's understanding of cultural and ideological development in the Liao Dynasty.

Sculptures that have been influenced by western civilizations can be seen in the Pagoda of the Kazuo Jingyan Temple, Chaoyang, Liaoning Province. For instance, there are evident traces that the sculptural structures of the physical shapes of Bodhisattvas have been influenced by sculptures in ancient Greek times. A pair of Bodhisattva sculptures located at the southwest of the first floor in the pagoda has been so perfectly made that they constitute a breakthrough from the previous complexity. Greater simplicity has been achieved by turning the draperies into several rhythmic lines that provide distinct feelings of architectural aesthetics. Such an aesthetic expression has been unprecedented in the history of Chinese artistic sculpture. It is a new sculptural language that has evolved amid cultural clashes, acceptance and creation between the Liao and western civilizations, which reveals the superb skills of artists especially in terms of realistic expressions. The exquisite and delicate depiction of drapery textures on the kasaya enables a feeling of translucency through the contrast of straight and wavy lines, bringing the liveliness into extremity. The control of sculptural space has been achieved in decisive, resumptive and trenchant rhythms. In particular, the strength, distance, and height variations of the contacting lines between the background wall and the sculptural form have been so artfully processed that the beauty in this is indescribable. Although the head and hands had varying degrees of damage, but it does not affect the value of the overall sculpture appreciation. The artistic style of the collision between the Khitan's national character and western civilization, as a historical witness of the ancient Silk Road civilization promoting cultural prosperity and development of the northern grassland, highlights the artists' pursuit and enterprising spirit of art ontology of beauty, and injects fresh blood into the sculpture art of Liao Dynasty.

It is fair to say that the King of heaven on the Huayanjing Pagoda in Hohhot, Inner Mongolia has achieved the best in his artistic expression. Although being incomplete, his physique is full of the artistic beauty of power which is just equal to that of the charm of the sculpture work from ancient Rome and the Renaissance period in Italy. The powerful and mighty physique seems to be able to touch your soul (Fig.12), thus being even fresher and more alive and vivid. Obviously, these sculpture thought and aesthetics have been influenced by the secular spirit of the works from ancient Rome period in the West, emphasizing scientific expression with perfect human-bodied proportion, athletic physique and exaggerated muscles, portrayed with human-like warmth. They look exactly like world-class statues of bodybuilders. Never before in the history of Chinese sculpture have people experienced such stunning sculptural art. They have broken through the deified and standardized images of the King of heaven in the traditional religion, which symbolize not only an intimidation to the vicious force but also the lofty responsibility of heroes in northern grassland nationality to protect the Great Liao.

The artists have sloughed off the traditionally deified King of heaven and created a heroic image to stand for the spirit of the northern grassland ethnicities. It seems as if people could actually hear the trembling sound of all those athletic muscles. The accurate grasp of the moving shapes, from the hips to the waist and from the legs to the feet, creates tremendous strength. Such body curve is so perfect, along with the waist-hip structural expression, thus reaching the acme of

Fig.12
The King of heaven
brick carving, on the first floor of the pagoda
Huayanjing Pagoda, Hohhot, Inner Mongolia

Fig.13
Gaius Octavius Augustus
marble carving
ancient Rome

perfection. The sculptured figure boasts strong visual effect with its magnificence, boldness and vigor, especially under such circumstance when there is a tower-ladder on the east side of the pagoda with which people can ascend to the first floor so as to get a better look at the body of the King of heaven on both sides in a shorter distance. At that moment people could not believe what they were appreciating was ancient Chinese sculptures. It's just like appreciating the sculpture work by masters from ancient Rome and the Renaissance period in Europe and surprised and captured by their charm completely (Fig.13). In an instant, the concept of the King of heaven has gone, only the art of sculpture is the only existence. The chunky molding, vivid carving marks, abstract physique, flowing lines, the extended space from the side view and the elastic muscles have made their physiques more stereoscopic and lively. The King of heaven on the north side of pagoda of first floor is a relatively well-preserved sculpture of the building except for its damaged head and arms. Artists have unleashed their enthusiasm to the fullest in this work through its powerful and mighty physique. The ribbon has been designed tactfully to fly over the figure and space, which has formed sharp contrast between the rigidity and flexibility, thus highlighting the beauty of the King of heaven's physiques. The linear structure created by decorative ribbon, shorts, waistband and amice has been integrated with the muscular power of the physique to form a beautiful harmony, thus constituting a perfect 'Symphony of Power' which is not only an appreciation of power but also a recognition of human wisdom. Therefore, this marks the birth of another sculpture form. The artists have created sculptures featuring the characteristics of grassland ethnics with a unique artistic style through realistic depiction, romantic idea and poetic passion, thus expressing the confidence, innovative spirit and artistic aesthetics of the artists in the Liao Dynasty.

If the King of heaven on the Huayanjing Pagoda is a praise of power, then Bodhisattva is a hymn to the beauty of physique.

The Bodhisattva on the first floor, on the southwest side of pagoda, although suffering severe damage with the loss of its head and arm, has been preserved relatively well. The charm in the art of the Bodhisattva sculpture has not diminished, but instead, it has the equal power of physical beauty as the King of heaven. Unlike the standardized and stereotyped shape of the traditional Bodhisattva, this is a whole new creation with human-like proportion and warmth. The sculpture is vivid, plump, magnificent, natural, stretched and flexible, which has paid great attention to the depiction of life forms in shaping the form. For example, in the intersection between the waist and the hip, the left leg has turned slightly inside with the center of gravity on the left foot and the right leg in relaxation, thus forming the dynamic body curve (Fig.14). The perfect curve has fully revealed the features of the whole body, and this elastic force is also commonly referred to as the 'beauty' of 'Venus' (Fig.15). Venus is an emblem of the Goddess of Beauty and Love with nine-headed body proportion so as to convey the 'beauty' of 'deity'. However, the Bodhisattva created by the artists of the Liao has eight-headed body, which has expressed the beauty of human physique in a scientific way. These works are the common aspiration for mankind in pursuing the ideal creation of beauty as well as an appreciation of beauty whether it's the beauty of deity or human. The artists in the Liao Dynasty have precisely captured the law of life motion which enables the motion curve to highlight the power of

artistic beauty. This style still continues in today's real life. For example, models in the world fashion shows have displayed their individual style of physique in a special manner which has been found and created by the artists from Liao Dynasty and later become the eternal power of the artistic beauty.

The sculpture of Daming Pagoda in Middle Capital of Liao Dynasty (Ningcheng County) of Inner Mongolia boosts a unique artistic form that combines painting and sculpture, with carved relief and colors. This artistic form has inherited the painted stone carving from the cave temple, but is slightly different. Due to the mixture painted of color and brick gray, the language of carved relief has faded while the effect of painting has been exaggerated. It is easy to see from the King of heaven on the northwest side of pagoda that greater attention has been paid to the spiritual similarity in the portray of their physique and motion, with exaggerated muscular strength and the detailed description of each facial expression, from the mimetic muscle to the eyes widened in shock or from the open mouth to the stomatic muscle. Such accurate and subtle expression is to warn the vicious force, and if one commits a crime, they will be shattered and brought to justice. In addition, the frown facial expression with both eyes staring straight ahead and tight lips of the King of heaven has conveyed an authoritative power over the forces of evil. The ornaments in the sculptural figure have been carved in great detail, including chaplet, necklace of jade and pearls, amice, heavens' cloths, blouse and belt, which is much like the white lines drawing in Chinese painting. In addition, the gray-brick natural color of the sculptural material and the faded vandyke red, azurite, mineral green, and mixed with a protective layer of white create similar effect of the meticulous brushwork rich in color in contemporary Chinese figure painting. Amazingly, it has conveyed the look of the King of heaven in a spiritually faithful manner with a little splash color effect. If such a unique visual effect applied to the creation of Chinese painting characters, it will produce unexpected shock. The charm of the aesthetic thoughts unleashed by the sculptural art of the Liao pagodas seems to be inexhaustible, thus offering people some fresh insights every time. This is differences between the sculpture art of Liao pagoda and other dynasties, and it is full of vitality.

The distinctive sculptural art of the Liao brick pagodas has displayed the whole process of the development history of sculpture in the Liao Dynasty, which resembles the ceaseless life in constant growth, thus being the legacy of the grassland ethnics of Liao Dynasty to empower today's sculptural art.

The brick pagoda sculpture in Liao Dynasty collected all the way has gone through the vicissitude of time, years and history. The precious artistic, cultural and historical heritage has created the earliest sculptural arts of realism in China through the evolution from traditional to modern times, from inheritance to development, from religion to secular life, from standardization to individualism, from deity to human, from absorption to creation, from plane to three-dimensional forms, from decoration to portrait, and from indoor to outdoor display. The unique culture and art of the grassland ethnics with its variety in forms, excellency in skills, rich and rare content in sculptural language and innovation in thinking and technology have greatly enriched the treasury of the history of Chinese cultural and artistic development.

The non-negligible incomplete sculptural art has also entered our era with a sense beyond the history itself. It possesses more vigor and vitality in the artistic aesthetics. The great creation of the artists in the Liao Dynasty has been integrated with history and time to create a new form of artistic language, being another cultural legacy left to us by the sculptural art of brick pagodas in the Liao. Such new language can do better in inspiring the imagination and passion for artistic creation. Modern artists have directly applied this incomplete aesthetic thinking and form in their works, which has produced a magic artistic and visual effect and thinking, and cannot be replaced by a complete image.

The sculptural art of brick pagodas in Liao Dynasty has proved that only through the absorption of advanced thinking and new breakthroughs in the creation can art prosper and flourish, and gain more national style and historical value.

All the above achievements and excavation have been gained by Professor Xu Bingkun during his compiling and summary of the accurate and detailed records written by Professor Bao Enli over the past thirty years. We will never accomplish all these without the guidance of Professor Xu Bingkun, which has demonstrated their admiration for the culture of the Northern grassland, their responsibility and dedication toward history. Therefore, it is possible for us to witness the glorious development of the culture of the northern grassland ethnics. If Professor Xu Bingkun has not insisted me accomplish the task, I would never encounter such a brilliant history, nor would I understand the abundance and splendor of the sculptural art in the Liao Dynasty. Our excavation and discovery also let people know that there once existed a glorious history of grassland ethnics in the history of the development of the Chinese nation. It is the stirrup culture created by grassland ethnics group that makes Europe enter into the modern society in advance, and at the same time, the European culture also promotes the development of grassland ethnics. The mutual learning between the culture of grassland ethnics and the European culture is the historical evidence in the field of brick sculptural arts. Therefore, it is not only a material evidence of the development of Chinese prairie culture, but also a material evidence of the development of world civilization. It deserves to be included into the world cultural heritage. Everyone has the responsibility to prevent the treasures of our country from vanishing anymore. We hope that the country can take actions to protect the grassland ethnics from becoming a pity in history. I would like to extend my sincere gratitude to Professor Xu Bingkun and Professor Bao Enli who have offered me such a precious opportunity and to my good friend Professor Zhang Shouguo who sacrificed his time to collect information with me all the way. My gratitude also goes to all the warm-hearted friends without whom we could never gather all the comprehensive and precious information about the sculptural art of brick pagodas in the Liao Dynasty. Everything we have done is worthwhile for discovering of a forgotten history in the development of Chinese Sculptural Art.

Zheng Bo
April 29, 2017 in Shenyang

砖 塔 雕 塑

BRICK-PAGODA SCULPTURE

塔 身

BODY OF THE PAGODA

1 ....
塔身
辽宁锦州广济寺塔

Pagoda body
Pagoda of Guangji Temple in Jinzhou,
Liaoning

2 ....
西面塔身
锦州广济寺塔

Pagoda body of the west side
Pagoda of Guangji Temple in Jinzhou

1

3
西面佛像头部
锦州广济寺塔
The head of Buddha on the west side
Pagoda of Guangji Temple in Jinzhou

4
西面佛像
锦州广济寺塔
Buddha on the west side
Pagoda of Guangji Temple in Jinzhou

4

5
北面佛像
锦州广济寺塔

Buddha on the north side
Pagoda of Guangji Temple in
Jinzhou

6
西北面佛像
锦州广济寺塔

Buddha on the northwest side
Pagoda of Guangji Temple in
Jinzhou

7
西北面佛像头部（正面）
锦州广济寺塔

**The head of Buddha on the
northwest side (front)**
Pagoda of Guangji Temple in
Jinzhou

8
西北面佛像头部（侧面）
锦州广济寺塔

**The head of Buddha on the
northwest side (side)**
Pagoda of Guangji Temple in
Jinzhou

6

113

南面塔身
辽宁朝阳凤凰山大宝塔

Pagoda body of the south side
Dabao Pagoda at Fenghuang
Mountain in Chaoyang, Liaoning

南面佛像局部
朝阳凤凰山大宝塔

Buddha on the south side (part)
Dabao Pagoda at Fenghuang
Mountain in Chaoyang

9

南面佛座·马
朝阳凤凰山大宝塔

Pedestal of Buddha on the south side · horses
Dabao Pagoda at Fenghuang Mountain in Chaoyang

12

东面佛像
辽宁朝阳凤凰山云接寺塔

Buddha on the east side
Pagoda of Yunjie Temple at Fenghuang
Mountain in Chaoyang, Liaoning

13

东面佛像头部
朝阳凤凰山云接寺塔

The head of Buddha on the east side
Pagoda of Yunjie Temple at Fenghuang
Mountain in Chaoyang

12

东面佛座·象
朝阳凤凰山云接寺塔

Pedestal of Buddha on the east side·elephants
Pagoda of Yunjie Temple at Fenghuang Mountain in Chaoyang

西面佛像
朝阳凤凰山云接寺塔

Buddha on the west side
Pagoda of Yunjie Temple at Fenghuang Mountain in Chaoyang

14

西面佛座・孔雀

朝阳凤凰山云接寺塔

Pedestal of Buddha on the west side · peacocks

Pagoda of Yunjie Temple at Fenghuang Mountain in Chaoyang

南面佛座・马

朝阳凤凰山云接寺塔

Pedestal of Buddha on the south side · horses

Pagoda of Yunjie Temple at Fenghuang Mountain in Chaoyang

16

18
南面佛像
辽宁朝阳北塔
Buddha on the south side
North Pagoda in Chaoyang,
Liaoning

19
南面佛座·孔雀
朝阳北塔
**Pedestal of Buddha on the
south side·peacocks**
North Pagoda in Chaoyang

19

20
东面佛像
朝阳北塔

Buddha on the east side
North Pagoda in Chaoyang

21
东面佛座·象
朝阳北塔

Pedestal of Buddha on the east side·elephants
North Pagoda in Chaoyang

21

22
....
南面佛座 · 马
朝阳北塔

**Pedestal of Buddha on
the south side · horse**
North Pagoda in Chaoyang

23
....
南面佛像
辽宁朝阳八棱观塔

Buddha on the south side
Balengguan Pagoda in Chaoyang,
Liaoning

22

东南面佛像
朝阳八棱观塔

Buddha on the southeast side
Balengguan Pagoda in Chaoyang

东面佛像
朝阳八棱观塔

Buddha on the east side
Balengguan Pagoda in Chaoyang

24

26 ....
东面佛像
内蒙古宁城大明塔

Buddha on the east side
Daming Pagoda in Ningcheng,
Inner Mongolia

27 ....
西北面佛像
宁城大明塔

Buddha on the northwest side
Daming Pagoda in Ningcheng

27

135

南面塔身左侧菩萨
辽宁海城金塔

Bodhisattva on the south side (left)
Jin Pagoda in Haicheng, Liaoning

南面塔身右侧菩萨
海城金塔

Bodhisattva on the south side (right)
Jin Pagoda in Haicheng

西南面塔身左侧菩萨
海城金塔

Bodhisattva on the southwest side (left)
Jin Pagoda in Haicheng

西南面塔身左侧菩萨手部
海城金塔

The left hand of Bodhisattva on the southwest side (left)
Jin Pagoda in Haicheng

西南面塔身右侧菩萨手部

海城金塔

**The left hand of Bodhisattva
on the southwest side (right)**

Jin Pagoda in Haicheng

西南面塔身右侧菩萨

海城金塔

**Bodhisattva on the southwest
side (right)**

Jin Pagoda in Haicheng

西北面塔身左侧菩萨
海城金塔

Bodhisattva on the northwest side (left)
Jin Pagoda in Haicheng

西北面塔身右侧菩萨
海城金塔

Bodhisattva on the northwest side (right)
Jin Pagoda in Haicheng

36
西南面塔身左侧菩萨
辽宁辽阳白塔

Bodhisattva on the southwest side (left)
White Pagoda in Liaoyang, Liaoning

37
西南面塔身右侧菩萨
辽阳白塔

Bodhisattva on the southwest side (right)
White Pagoda in Liaoyang

西北面塔身左侧菩萨
辽阳白塔

Bodhisattva on the northwest side (left)
White Pagoda in Liaoyang

39

西北面塔身右侧菩萨
辽阳白塔

Bodhisattva on the northwest side (right)
White Pagoda in Liaoyang

40

北面塔身左侧菩萨
辽阳白塔

Bodhisattva on the north side (left)
White Pagoda in Liaoyang

41

北面塔身右侧菩萨
辽阳白塔

Bodhisattva on the north side (right)
White Pagoda in Liaoyang

42
北面塔身左侧菩萨局部
辽阳白塔

Part of Bodhisattva on
the north side (left)
White Pagoda in Liaoyang

43
北面塔身右侧菩萨局部
辽阳白塔

Part of Bodhisattva on
the north side (right)
White Pagoda in Liaoyang

南面塔身右侧菩萨局部
辽阳白塔

Part of Bodhisattva on the south side (right)
White Pagoda in Liaoyang

南面塔身右侧菩萨
辽阳白塔

Bodhisattva on the south side (right)
White Pagoda in Liaoyang

北面塔身左侧菩萨
辽宁锦州广济寺塔

Bodhisattva on the north side (left)
Pagoda of Guangji Temple in Jinzhou, Liaoning

北面塔身右侧菩萨
锦州广济寺塔

Bodhisattva on the north side (right)
Pagoda of Guangji Temple in Jinzhou

48
西面塔身左侧菩萨
锦州广济寺塔

Bodhisattva on the west side (left)
Pagoda of Guangji Temple in Jinzhou

49
西面塔身左侧菩萨局部
锦州广济寺塔

Part of Bodhisattva on the west side (left)
Pagoda of Guangji Temple in Jinzhou

西面塔身右侧菩萨
锦州广济寺塔
Bodhisattva on the west side (right)
Pagoda of Guangji Temple in Jinzhou

西面塔身右侧菩萨头部
锦州广济寺塔
The head of Bodhisattva on the west side (right)
Pagoda of Guangji Temple in Jinzhou

南面塔身左侧菩萨
辽宁朝阳北塔

**Bodhisattva on the south
side (left)**
North Pagoda in Chaoyang,
Liaoning

南面塔身右侧菩萨
朝阳北塔

**Bodhisattva on the south
side (right)**
North Pagoda in Chaoyang

54

二层东北面塔身菩萨（正面）
辽宁喀左精严禅寺塔

**Bodhisattvas on the northeast side
of pagoda of the second floor (front)**
Pagoda of Jingyan Temple in Kazuo,
Liaoning

二层东北面塔身菩萨（侧面）
喀左精严禅寺塔

**Bodhisattvas on the northeast side
of pagoda of the second floor (side)**
Pagoda of Jingyan Temple in Kazuo

56

一层东南面塔身菩萨
喀左精严禅寺塔

Bodhisattvas on the southeast
side of pagoda of the first floor
Pagoda of Jingyan Temple in Kazuo

一层东南面塔身左侧菩萨局部
喀左精严禅寺塔

Part of Bodhisattva (left) on the
southeast side of pagoda of the
first floor
Pagoda of Jingyan Temple in Kazuo

58

二层东南面塔身菩萨
喀左精严禅寺塔

**Bodhisattvas on the southeast side
of pagoda of the second floor**
Pagoda of Jingyan Temple in Kazuo

二层东南面塔身左侧菩萨
喀左精严禅寺塔

**Bodhisattva (left) on the southeast
side of pagoda of the second floor**
Pagoda of Jingyan Temple in Kazuo

60

二层东南面塔身右侧菩萨
喀左精严禅寺塔

**Bodhisattva (right) on the southeast
side of pagoda of the second floor**
Pagoda of Jingyan Temple in Kazuo

二层西南面塔身菩萨（侧面）
喀左精严禅寺塔

**Bodhisattvas (side) on the southwest
side of pagoda of the second floor**
Pagoda of Jingyan Temple in Kazuo

62

一层西北面塔身菩萨
喀左精严禅寺塔

Bodhisattvas on the northwest side of pagoda of the first floor
Pagoda of Jingyan Temple in Kazuo

一层西北面塔身左侧菩萨
喀左精严禅寺塔

Bodhisattva (left) on the northwest side of pagoda of the first floor
Pagoda of Jingyan Temple in Kazuo

64

一层西北面塔身右侧菩萨
喀左精严禅寺塔

**Bodhisattva (right) on the northwest
side of pagoda of the first floor**
Pagoda of Jingyan Temple in Kazuo

二层西北面塔身菩萨
喀左精严禅寺塔

**Bodhisattvas on the northwest side
of pagoda of the second floor**
Pagoda of Jingyan Temple in Kazuo

66

一层西南面塔身菩萨（正面）
喀左精严禅寺塔

**Bodhisattvas (front) on the southwest
side of pagoda of the first floor**
Pagoda of Jingyan Temple in Kazuo

一层西南面塔身左侧菩萨
喀左精严禅寺塔

Bodhisattva (left) on the southwest side of pagoda of the first floor
Pagoda of Jingyan Temple in Kazuo

70
一层西南面塔身右侧菩萨
喀左精严禅寺塔

Bodhisattva (right) on the southwest side of pagoda of the first floor
Pagoda of Jingyan Temple in Kazuo

71
一层西南面塔身菩萨（侧面）
喀左精严禅寺塔

**Bodhisattvas (side) on
the southwest side of
pagoda of the first floor**
Pagoda of Jingyan Temple in
Kazuo

72
东面塔身左侧菩萨
内蒙古宁城大明塔

**Bodhisattva (left) on the
east side**
Daming Pagoda in Ningcheng,
Inner Mongolia

东面塔身右侧菩萨
宁城大明塔

Bodhisattva on the east side (right)
Daming Pagoda in Ningcheng

东面塔身右侧菩萨局部
宁城大明塔

Part of Bodhisattva on the east side (right)
Daming Pagoda in Ningcheng

75
南面塔身左侧菩萨
宁城大明塔

Bodhisattva on the south side (left)
Daming Pagoda in Ningcheng

76
南面塔身右侧菩萨
宁城大明塔

Bodhisattva on the south side (right)
Daming Pagoda in Ningcheng

二层西南面塔身之一
内蒙古呼和浩特华严经塔

二层西南面塔身之二
呼和浩特华严经塔

Pagoda body of the southwest side of the second floor (first)
Huayanjing Pagoda in Hohhot, Inner Mongolia

Pagoda body of the southwest side of the second floor (second)
Huayanjing Pagoda in Hohhot

77

二层西南面塔身左侧菩萨头部
呼和浩特华严经塔

The head of Bodhisattva (left) on the southwest
side of pagoda of the second floor
Huayanjing Pagoda in Hohhot

二层西南面塔身左侧菩萨（正面）
呼和浩特华严经塔

Bodhisattva (left) on the southwest side of pagoda
of the second floor (front)
Huayanjing Pagoda in Hohhot

81
....
二层东南面塔身右侧菩萨头部
呼和浩特华严经塔
The head of Bodhisattva (right) on the southeast side of pagoda of the second floor
Huayanjing Pagoda in Hohhot

82
....
二层东南面塔身右侧菩萨（侧面）
呼和浩特华严经塔
Bodhisattva (right) on the southeast side of pagoda of the second floor (side)
Huayanjing Pagoda in Hohhot

83
....
二层东南面塔身右侧菩萨（正面）
呼和浩特华严经塔
Bodhisattva (right) on the southeast side of pagoda of the second floor (front)
Huayanjing Pagoda in Hohhot

82

83

一层东北面塔身左侧菩萨
呼和浩特华严经塔
**Bodhisattva (left) on the northeast
side of pagoda of the first floor**
Huayanjing Pagoda in Hohhot

一层东北面塔身右侧菩萨
呼和浩特华严经塔
**Bodhisattva (right) on the northeast
side of pagoda of the first floor**
Huayanjing Pagoda in Hohhot

84

85

86
一层西南面塔身左侧菩萨
呼和浩特华严经塔
Bodhisattva (left) on the southwest side of pagoda of the first floor
Huayanjing Pagoda in Hohhot

87
一层东南面塔身右侧菩萨
呼和浩特华严经塔
Bodhisattva (right) on the southeast side of pagoda of the first floor
Huayanjing Pagoda in Hohhot

87

一层西北面塔身左侧菩萨
呼和浩特华严经塔
Bodhisattva (left) on the northwest side of pagoda of the first floor
Huayanjing Pagoda in Hohhot

一层西北面塔身右侧菩萨
呼和浩特华严经塔
Bodhisattva (right) on the northwest side of pagoda of the first floor
Huayanjing Pagoda in Hohhot

北面塔身飞天
辽宁海城金塔

Flying Apsaras on the north side
Jin Pagoda in Haicheng, Liaoning

东面塔身飞天
海城金塔

Flying Apsaras on the east side
Jin Pagoda in Haicheng

90

94
....
东面塔身左侧飞天
海城金塔

**Flying Apsaras on
the east side (left)**
Jin Pagoda in Haicheng

95
....
东面塔身右侧飞天
海城金塔

**Flying Apsaras on
the east side (right)**
Jin Pagoda in Haicheng

94

96

东南面塔身飞天
辽宁辽阳白塔

Flying Apsaras on the southeast side
White Pagoda in Liaoyang, Liaoning

南面塔身飞天
辽阳白塔

Flying Apsaras on the south side
White Pagoda in Liaoyang

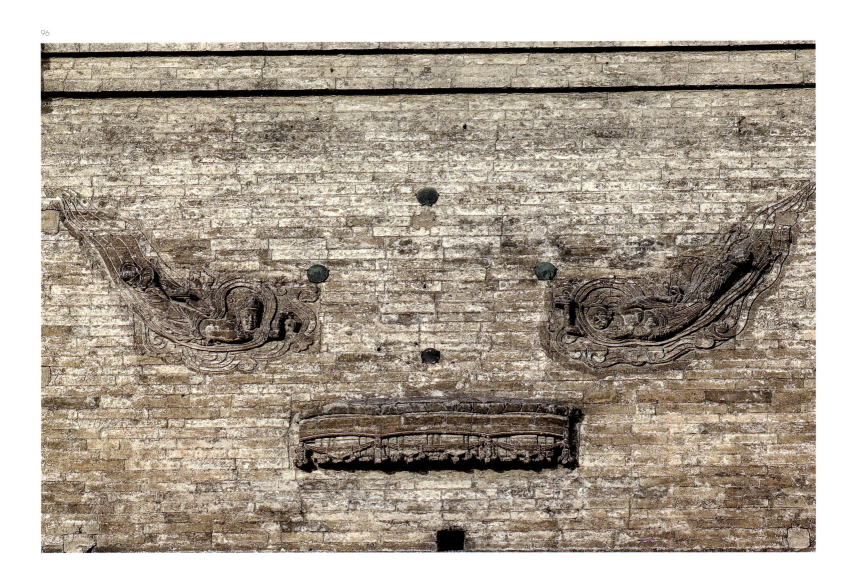

东南面塔身左侧飞天
辽阳白塔
**Flying Apsaras on the
southeast side (left)**
White Pagoda in Liaoyang

东南面塔身右侧飞天
辽阳白塔
**Flying Apsaras on the
southeast side (right)**
White Pagoda in Liaoyang

98

100

南面塔身左侧飞天
辽阳白塔

**Flying Apsaras on the
south side (left)**
White Pagoda in Liaoyang

101

南面塔身右侧飞天
辽阳白塔

**Flying Apsaras on the
south side (right)**
White Pagoda in Liaoyang

101

102

西北面塔身飞天
辽阳白塔

**Flying Apsaras on the
northwest side**
White Pagoda in Liaoyang

103

东面塔身飞天
辽阳白塔

**Flying Apsaras on the
east side**
White Pagoda in Liaoyang

102

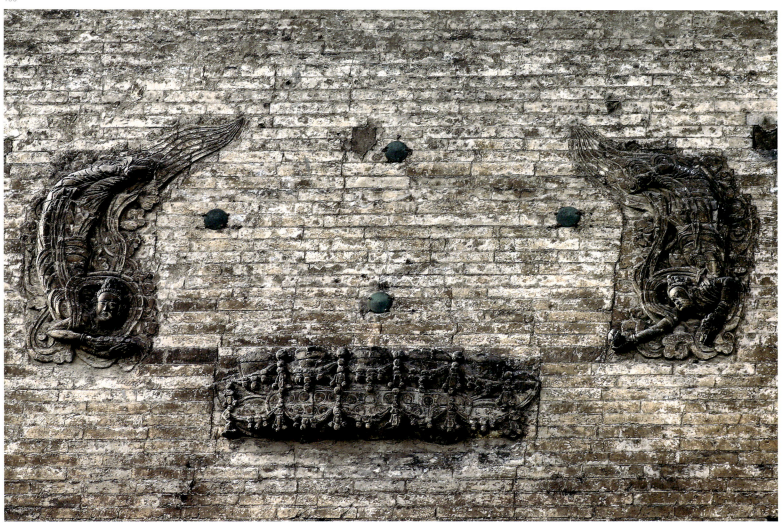

西北面塔身左侧飞天
辽阳白塔

**Flying Apsaras on the
northwest side (left)**
White Pagoda in Liaoyang

西北面塔身右侧飞天
辽阳白塔

**Flying Apsaras on the
northwest side (right)**
White Pagoda in Liaoyang

东面塔身左侧飞天
辽阳白塔

**Flying Apsaras on the
east side (left)**
White Pagoda in Liaoyang

东面塔身右侧飞天
辽阳白塔

**Flying Apsaras on the
east side (right)**
White Pagoda in Liaoyang

104

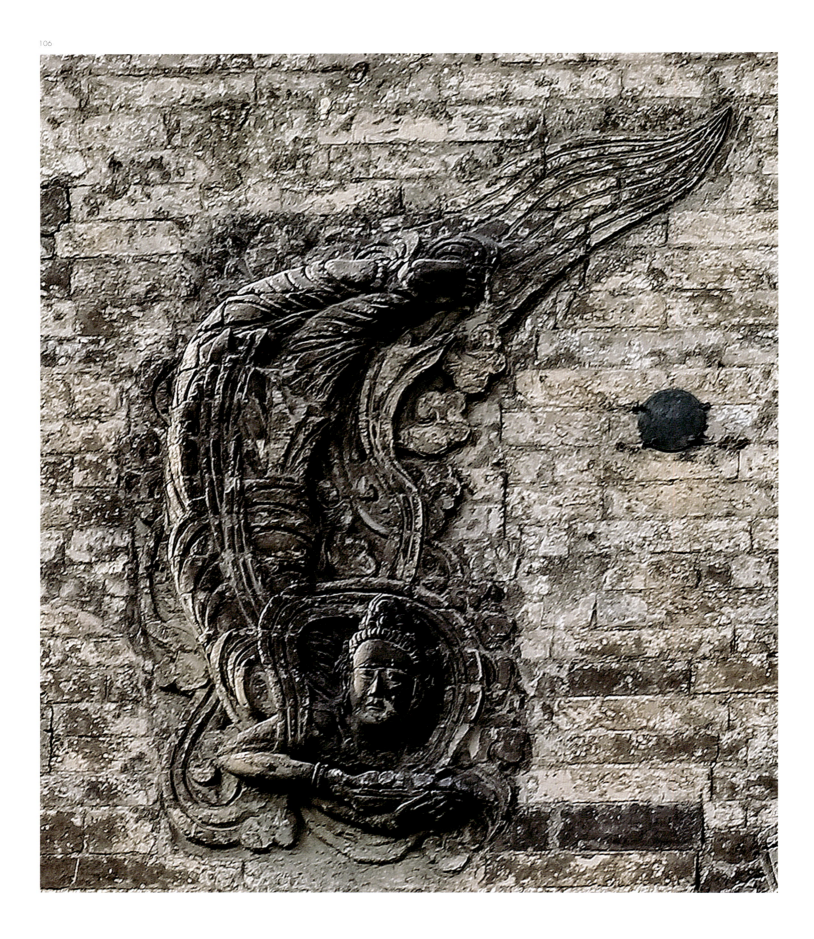

西北面塔身飞天
辽宁锦州广济寺塔

Flying Apsaras on the northwest side
Pagoda of Guangji Temple in Jinzhou, Liaoning

西北面塔身左侧飞天
锦州广济寺塔

Flying Apsaras on the northwest side (left)
Pagoda of Guangji Temple in Jinzhou

西北面塔身右侧飞天
锦州广济寺塔

Flying Apsaras on the northwest side (right)
Pagoda of Guangji Temple in Jinzhou

西面塔身左侧飞天
锦州广济寺塔

**Flying Apsaras on the
west side (left)**
Pagoda of Guangji Temple
in Jinzhou

西面塔身右侧飞天
锦州广济寺塔

**Flying Apsaras on the
west side (right)**
Pagoda of Guangji Temple
in Jinzhou

南面塔身左侧飞天
辽宁朝阳凤凰山大宝塔

Flying Apsaras on the south side (left)
Dabao Pagoda at Fenghuang
Mountain in Chaoyang, Liaoning

南面塔身右侧飞天
朝阳凤凰山大宝塔

Flying Apsaras on the south side (right)
Dabao Pagoda at Fenghuang
Mountain in Chaoyang

113

西面塔身左侧飞天
辽宁朝阳凤凰山云接寺塔

Flying Apsaras on the west side (left)
Pagoda of Yunjie Temple at Fenghuang
Mountain in Chaoyang, Liaoning

西面塔身右侧飞天
朝阳凤凰山云接寺塔

Flying Apsaras on the west side (right)
Pagoda of Yunjie Temple at Fenghuang
Mountain in Chaoyang

115

东面塔身左侧飞天
辽宁朝阳北塔
Flying Apsaras on the east side (left)
North Pagoda in Chaoyang, Liaoning

东面塔身右侧飞天
朝阳北塔
Flying Apsaras on the east side (right)
North Pagoda in Chaoyang

117

119
南面塔身左侧飞天
朝阳北塔

Flying Apsaras on the south side (left)
North Pagoda in Chaoyang

120
南面塔身右侧飞天
朝阳北塔

Flying Apsaras on the south side (right)
North Pagoda in Chaoyang

119

121

东面塔身左侧飞天
内蒙古宁城大明塔

Flying Apsaras on the east side (left)
Daming Pagoda in Ningcheng, Inner Mongolia

122

东面塔身右侧飞天
宁城大明塔

Flying Apsaras on the east side (right)
Daming Pagoda in Ningcheng

121

南面塔身左侧飞天
宁城大明塔
Flying Apsaras on the south side (left)
Daming Pagoda in Ningcheng

南面塔身右侧飞天
宁城大明塔
Flying Apsaras on the south side (right)
Daming Pagoda in Ningcheng

123

西面塔身左侧飞天
宁城大明塔

Flying Apsaras on the west
side (left)
Daming Pagoda in Ningcheng

西面塔身右侧飞天
宁城大明塔

Flying Apsaras on the west
side (right)
Daming Pagoda in Ningcheng

125

126

一层西北面塔身
内蒙古巴林右旗庆州白塔

Pagoda body of the northwest side of the first floor
Qingzhou White Pagoda in Balin Youqi, Inner Mongolia

一层西南面塔身
巴林右旗庆州白塔

Pagoda body of the southwest side of the first floor
Qingzhou White Pagoda in Balin Youqi

127

129
一层西北面塔身飞天（迦陵频伽）
巴林右旗庆州白塔
Flying Apsaras (Kalavinka) on the northwest side of pagoda of the first floor
Qingzhou White Pagoda in Balin Youqi

130
一层西南面塔身飞天（迦陵频伽）
巴林右旗庆州白塔
Flying Apsaras (Kalavinka) on the southwest side of pagoda of the first floor
Qingzhou White Pagoda in Balin Youqi

131
塔身飞天之一
内蒙古巴林左旗南塔
巴林左旗博物馆藏

Flying Apsaras (first)
South Pagoda in Balin Zuoqi, Inner
Mongolia, in the collections of the
Balin Zuoqi Museum

132
塔身飞天之二
巴林左旗南塔
巴林左旗博物馆藏

Flying Apsaras (second)
South Pagoda in Balin Zuoqi, in
the collections of the Balin Zuoqi
Museum

133

南面塔身左侧飞天
北京房山云居寺北塔
**Flying Apsaras on the south
side (left)**
North Pagoda of Yunju Temple in
Fangshan, Beijing

134

南面塔身右侧飞天
房山云居寺北塔
**Flying Apsaras on the south
side (right)**
North Pagoda of Yunju Temple in
Fangshan

二层塔身
辽宁喀左精严禅寺塔

Part of pagoda of the second floor
Pagoda of Jingyan Temple in Kazuo,
Liaoning

二层南面塔身天王
喀左精严禅寺塔

**The Kings of heaven on the south
side of pagoda of the second floor**
Pagoda of Jingyan Temple in Kazuo

135

二层南面塔身左侧天王
喀左精严禅寺塔
The King of heaven (left) on the south
side of pagoda of the second floor
Pagoda of Jingyan Temple in Kazuo

二层南面塔身右侧天王
喀左精严禅寺塔
The King of heaven (right) on the south
side of pagoda of the second floor
Pagoda of Jingyan Temple in Kazuo

137

138

一层西面塔身天王
喀左精严禅寺塔

The Kings of heaven on the west
side of pagoda of the first floor
Pagoda of Jingyan Temple in Kazuo

一层西面塔身左侧天王头部
喀左精严禅寺塔

The head of the King of heaven (left) on
the west side of pagoda of the first floor
Pagoda of Jingyan Temple in Kazuo

139

一层西面塔身左侧天王
喀左精严禅寺塔

The King of heaven (left) on the west
side of pagoda of the first floor
Pagoda of Jingyan Temple in Kazuo

一层西面塔身右侧天王
喀左精严禅寺塔

The King of heaven (right) on the west
side of pagoda of the first floor
Pagoda of Jingyan Temple in Kazuo

141

143

一层西面塔身右侧天王头部
喀左精严禅寺塔

The head of the King of heaven (right)
on the west side of pagoda of the first
floor

Pagoda of Jingyan Temple in Kazuo

144

二层西面塔身天王
喀左精严禅寺塔

The Kings of heaven on the west
side of pagoda of the second floor

Pagoda of Jingyan Temple in Kazuo

144

145
二层西面塔身左侧天王
喀左精严禅寺塔
The King of heaven (left) on the west side of pagoda of the second floor
Pagoda of Jingyan Temple in Kazuo

146
二层西面塔身右侧天王
喀左精严禅寺塔
The King of heaven (right) on the west side of pagoda of the second floor
Pagoda of Jingyan Temple in Kazuo

147

二层西面塔身右侧天王头部
喀左精严禅寺塔

The head of the King of heaven
(right) on the west side of pagoda
of the second floor

Pagoda of Jingyan Temple in Kazuo

148

二层西面塔身右侧天王（侧面）
喀左精严禅寺塔

The King of heaven (right) on the
west side of pagoda of the second
floor (side)

Pagoda of Jingyan Temple in Kazuo

148

二层东面塔身左侧天王
喀左精严禅寺塔

The King of heaven (left) on
the east side of pagoda of the
second floor
Pagoda of Jingyan Temple in Kazuo

二层东面塔身右侧天王
喀左精严禅寺塔

The King of heaven (right) on
the east side of pagoda of the
second floor
Pagoda of Jingyan Temple in Kazuo

149

二层北面塔身
喀左精严禅寺塔

Pagoda body of the north side of the second floor
Pagoda of Jingyan Temple in Kazuo

南面、东南面、东面塔身（左→右）
内蒙古宁城大明塔

Pagoda body on the south, southeast, east side (from left to right)
Daming Pagoda in Ningcheng, Inner Mongolia

西北面塔身
宁城大明塔

Pagoda body on the northwest side
Daming Pagoda in Ningcheng

152

154
东南面塔身左侧天王
宁城大明塔
**The King of heaven on
the southeast side (left)**
Daming Pagoda in Ningcheng

155
东南面塔身右侧天王
宁城大明塔
**The King of heaven on
the southeast side (right)**
Daming Pagoda in Ningcheng

156
西南面塔身左侧天王
宁城大明塔
The King of heaven on the southwest side (left)
Daming Pagoda in Ningcheng

157
西南面塔身右侧天王
宁城大明塔
The King of heaven on the southwest side (right)
Daming Pagoda in Ningcheng

158
西北面塔身左侧天王
宁城大明塔
The King of heaven on the northwest side (left)
Daming Pagoda in Ningcheng

159
西北面塔身左侧天王头部
宁城大明塔
The head of the King of heaven on the northwest side (left)
Daming Pagoda in Ningcheng

西北面塔身右侧天王头部
宁城大明塔

**The head of the King of heaven on the
northwest side (right)**
Daming Pagoda in Ningcheng

西北面塔身右侧天王
宁城大明塔

**The King of heaven on the northwest
side (right)**
Daming Pagoda in Ningcheng

162
东北面塔身左侧天王
宁城大明塔
The King of heaven on the northeast side (left)
Daming Pagoda in Ningcheng

163
东北面塔身右侧天王
宁城大明塔
The King of heaven on the northeast side (right)
Daming Pagoda in Ningcheng

一层西面塔身左侧天王
内蒙古巴林右旗庆州白塔

**The King of heaven (left) on the west
side of pagoda of the first floor**
Qingzhou White Pagoda in Balin Youqi,
Inner Mongolia

一层西面塔身右侧天王
巴林右旗庆州白塔

**The King of heaven (right) on the west
side of pagoda of the first floor**
Qingzhou White Pagoda in Balin Youqi

二层南面塔身
内蒙古呼和浩特华严经塔

Pagoda body of the south side of the second floor
Huayanjing Pagoda in Hohhot, Inner Mongolia

167

二层南面塔身左侧天王
呼和浩特华严经塔

The King of heaven (left) on the south side of pagoda of the second floor
Huayanjing Pagoda in Hohhot

166

二层南面塔身右侧天王
呼和浩特华严经塔

The King of heaven (right)
on the south side of pagoda
of the second floor
Huayanjing Pagoda in Hohhot

二层南面塔身右侧天王局部
呼和浩特华严经塔

Part of the King of heaven (right)
on the south side of pagoda of
the second floor
Huayanjing Pagoda in Hohhot

168

170
二层西面塔身右侧天王
呼和浩特华严经塔
**The King of heaven (right)
on the west side of pagoda
of the second floor**
Huayanjing Pagoda in Hohhot

171
二层西面塔身左侧天王
呼和浩特华严经塔
**The King of heaven (left)
on the west side of pagoda
of the second floor**
Huayanjing Pagoda in Hohhot

二层北面塔身左侧天王
呼和浩特华严经塔

The King of heaven (left) on
the north side of pagoda of
the second floor
Huayanjing Pagoda in Hohhot

二层北面塔身右侧天王
呼和浩特华严经塔

The King of heaven (right)
on the north side of pagoda
of the second floor
Huayanjing Pagoda in Hohhot

一层北面塔身
呼和浩特华严经塔

**Pagoda body on the north
side of the first floor**
Huayanjing Pagoda in Hohhot

一层西面塔身
呼和浩特华严经塔

**Pagoda body on the west side
of the first floor**
Huayanjing Pagoda in Hohhot

174

176 ....
一层北面塔身左侧天王
呼和浩特华严经塔
The King of heaven (left) on the north side of pagoda of the first floor
Huayanjing Pagoda in Hohhot

177 ....
一层北面塔身右侧天王
呼和浩特华严经塔
The King of heaven (right) on the north side of pagoda of the first floor
Huayanjing Pagoda in Hohhot

一层西面塔身左侧天王
呼和浩特华严经塔
**The King of heaven (left)
on the west side of pagoda
of the first floor**
Huayanjing Pagoda in Hohhot

一层西面塔身右侧天王
呼和浩特华严经塔
**The King of heaven (right)
on the west side of pagoda
of the first floor**
Huayanjing Pagoda in Hohhot

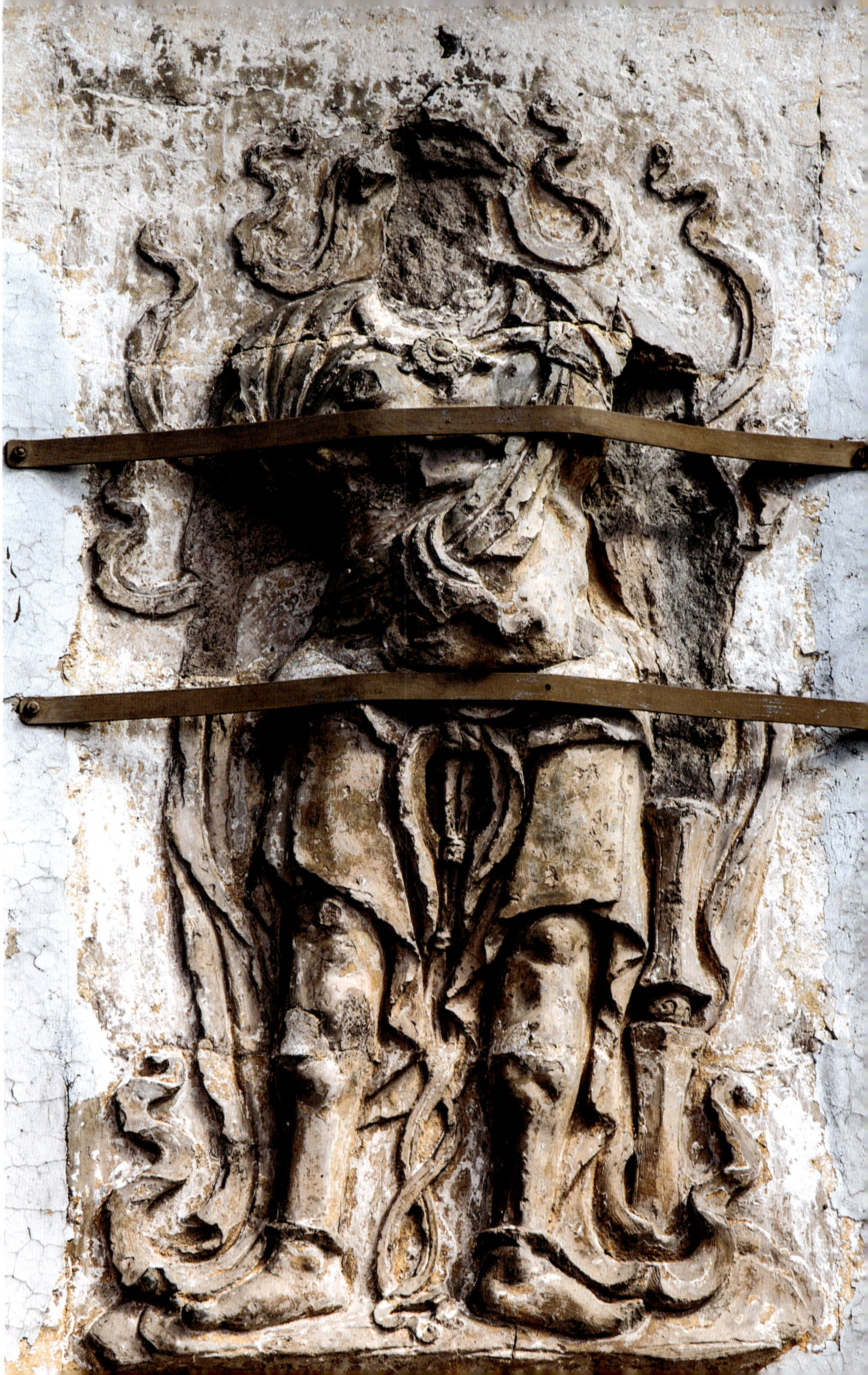

180

一层东南面塔身左侧天王

呼和浩特华严经塔

The King of heaven (left) on the southeast side of pagoda of the first floor

Huayanjing Pagoda in Hohhot

181

一层南面塔身右侧天王

呼和浩特华严经塔

The King of heaven (right) on the south side of pagoda of the first floor

Huayanjing Pagoda in Hohhot

180

须弥座

SUMERU PODIUM

塔身和二层须弥座
辽宁海城金塔

Pagoda body and two-layered sumeru podium
Jin Pagoda in Haicheng, Liaoning

二层须弥座
海城金塔

Two-layered sumeru podium
Jin Pagoda in Haicheng

182

184
一层须弥座东北面北角负塔力士
海城金塔
Strong man (north corner) on
the northeast side of sumeru
podium of the first floor
Jin Pagoda in Haicheng

185
一层须弥座东南面南角负塔力士
海城金塔
Strong man (south corner) on
the southeast side of sumeru
podium of the first floor
Jin Pagoda in Haicheng

184

186
一层须弥座东北面东角负塔力士头部
海城金塔
The head of strong man (east corner) on the northeast side of sumeru podium of the first floor
Jin Pagoda in Haicheng

187
一层须弥座东北面东角负塔力士
海城金塔
Strong man (east corner) on the northeast side of sumeru podium of the first floor
Jin Pagoda in Haicheng

187

188
二层须弥座东北面北角负塔力士
海城金塔
Strong man (north corner) on the northeast side of sumeru podium of the second floor
Jin Pagoda in Haicheng

189
东北角负塔力士
辽宁朝阳凤凰山云接寺塔
Strong man at the northeast corner
Pagoda of Yunjie Temple at Fenghuang Mountain in Chaoyang, Liaoning

190
西南角负塔力士（西侧）
朝阳凤凰山云接寺塔
Strong man at the southwest corner (west)
Pagoda of Yunjie Temple at Fenghuang Mountain in Chaoyang

189

190

191
西南角负塔力士（正面）
朝阳凤凰山云接寺塔
Strong man at the southwest corner (front)
Pagoda of Yunjie Temple at Fenghuang Mountain in Chaoyang

192
西北角负塔力士
朝阳凤凰山云接寺塔
Strong man at the northwest corner
Pagoda of Yunjie Temple at Fenghuang Mountain in Chaoyang

193
西南角负塔力士（南侧）
朝阳凤凰山云接寺塔
Strong man at the southwest corner (south)
Pagoda of Yunjie Temple at Fenghuang Mountain in Chaoyang

192

193

须弥座西面假门
辽宁朝阳北塔

Blank door on the west side of sumeru podium
North Pagoda in Chaoyang, Liaoning

须弥座西面假门右侧天王
朝阳北塔

The King of heaven on the right side of blank door of sumeru podium (west)
North Pagoda in Chaoyang

194

196
须弥座东面假门左侧天王
朝阳北塔
The King of heaven on the
left side of blank door of
sumeru podium (east)
North Pagoda in Chaoyang

197
须弥座东面假门
朝阳北塔
Blank door on the east side
of sumeru podium
North Pagoda in Chaoyang

196

须弥座西面西北角负塔力士

辽宁喀左精严禅寺塔

**Strong man at the northwest corner
of sumeru podium (west)**

Pagoda of Jingyan Temple in Kazuo, Liaoning

须弥座南面东南角负塔力士

北京房山云居寺北塔

**Strong man at the Southeast corner
of sumeru podium (south)**

North Pagoda of Yunju Temple in Fangshan,
Beijing

198

须弥座北面东北角负塔力士
房山云居寺北塔
Strong man at the northeast corner
of sumeru podium (north)
North Pagoda of Yunju Temple in Fangshan

须弥座北面西北角负塔力士
房山云居寺北塔
Strong man at the northwest corner
of sumeru podium (north)
North Pagoda of Yunju Temple in Fangshan

200

须弥座东南面乐人之一
辽宁海城金塔

**Musician on the southeast side
of sumeru podium (first)**

Jin Pagoda in Haicheng, Liaoning

须弥座东南面乐人之二
海城金塔

**Musician on the southeast side
of sumeru podium (second)**

Jin Pagoda in Haicheng

202

204 ....

须弥座西面乐人之一

辽宁朝阳北塔

**Musicians on the west side
of sumeru podium (first)**

North Pagoda in Chaoyang,
Liaoning

205 ....

须弥座西面乐人之二

朝阳北塔

**Musicians on the west side
of sumeru podium (second)**

North Pagoda in Chaoyang

204

206
须弥座北面舞人之一
北京房山云居寺北塔
Dancer on the north side of sumeru podium (first)
North Pagoda of Yunju Temple in Fangshan, Beijing

207
须弥座北面舞人之二
房山云居寺北塔
Dancer on the north side of sumeru podium (second)
North Pagoda of Yunju Temple in Fangshan

206

须弥座东北面舞乐人
房山云居寺北塔

**Dancer and musician on the northeast
side of sumeru podium**
North Pagoda of Yunju Temple in Fangshan

须弥座东南面乐人
房山云居寺北塔

**Musician on the southeast side
of sumeru podium**
North Pagoda of Yunju Temple in
Fangshan

208

须弥座东北面乐人
房山云居寺北塔
**Musician on the northeast side
of sumeru podium**
North Pagoda of Yunju Temple in
Fangshan

须弥座西面乐人之一
房山云居寺北塔
**Musician on the west side
of sumeru podium (first)**
North Pagoda of Yunju Temple
in Fangshan

210

须弥座西面乐人之二
房山云居寺北塔
**Musician on the west side of
sumeru podium (second)**
*North Pagoda of Yunju Temple in
Fangshan*

须弥座西面舞人
房山云居寺北塔
**Dancer on the west side of
sumeru podium**
*North Pagoda of Yunju Temple
in Fangshan*

212

214

须弥座西南面乐人
房山云居寺北塔

**Musician on the southwest
side of sumeru podium**
North Pagoda of Yunju Temple in
Fangshan

215

须弥座北面舞乐人
房山云居寺北塔

**Dancers and Musicians on the
north side of sumeru podium**
North Pagoda of Yunju Temple in
Fangshan

214

215

须弥座北面乐人
房山云居寺北塔

**Musician on the north side of
sumeru podium**
North Pagoda of Yunju Temple in
Fangshan

须弥座北面舞人之三
房山云居寺北塔

**Dancer on the north side
of sumeru podium (third)**
North Pagoda of Yunju Temple in
Fangshan

216

须弥座南面供养人
辽宁朝阳凤凰山大宝塔
Buddhist donators on the south side of sumeru podium
Dabao Pagoda at Fenghuang Mountain in Chaoyang, Liaoning

须弥座南面供养人
北京房山云居寺北塔
Buddhist donator on the south side of sumeru podium
North Pagoda of Yunju Temple in Fangshan, Beijing

218

Buddhist donators on the south side of sumeru podium
Dabao Pagoda at Fenghuang Mountain in Chaoyang, Liaoning

须弥座西南面迦陵频伽之一
房山云居寺北塔

**Kalavinka on the southwest side
of sumeru podium (first)**
North Pagoda of Yunju Temple in
Fangshan

须弥座西南面迦陵频伽之二
房山云居寺北塔

**Kalavinka on the southwest side
of sumeru podium (second)**
North Pagoda of Yunju Temple in
Fangshan

220

须弥座北面供养人之一
辽宁海城金塔
Buddhist donator on the north
side of sumeru podium (first)
Jin Pagoda in Haicheng, Liaoning

须弥座北面供养人之二
海城金塔
Buddhist donators on the north
side of sumeru podium (second)
Jin Pagoda in Haicheng

222

须弥座东北面供养人之一
海城金塔

**Buddhist donator on the northeast
side of sumeru podium (first)**

Jin Pagoda in Haicheng

须弥座东北面供养人之二
海城金塔

**Buddhist donator on the northeast
side of sumeru podium (second)**

Jin Pagoda in Haicheng

224

226
须弥座东北面供养人之三
海城金塔
Buddhist donator on the northeast
side of sumeru podium (third)
Jin Pagoda in Haicheng

227
须弥座西南面供养人
海城金塔
Buddhist donator on the southwest
side of sumeru podium
Jin Pagoda in Haicheng

226

须弥座西南面供养人之一
北京房山云居寺北塔
Buddhist donator on the southwest side of sumeru podium (first)
North Pagoda of Yunju Temple in Fangshan, Beijing

须弥座西南面供养人之二
房山云居寺北塔
Buddhist donator on the southwest side of sumeru podium (second)
North Pagoda of Yunju Temple in Fangshan

228

230

230
须弥座乐人与供养人
房山云居寺北塔
Buddhist donators and musician on the sumeru podium
North Pagoda of Yunju Temple in Fangshan

231
须弥座东南面骑动物者
房山云居寺北塔
Animal rider on the southeast side of sumeru podium
North Pagoda of Yunju Temple in Fangshan

须弥座南面供养人之一
房山云居寺北塔
**Buddhist donators on the south side
of sumeru podium (first)**
North Pagoda of Yunju Temple in Fangshan

须弥座南面供养人之二
房山云居寺北塔
**Buddhist donators on the south side
of sumeru podium (second)**
North Pagoda of Yunju Temple in Fangshan

232

233

须弥座西面礼佛图
辽宁喀左精严禅寺塔

Picture of praying to Buddha on the west side of sumeru podium
Pagoda of Jingyan Temple in Kazuo, Liaoning

234

须弥座东南面礼佛图
喀左精严禅寺塔
Picture of praying to Buddha on the southeast side of sumeru podium
Pagoda of Jingyan Temple in Kazuo

235

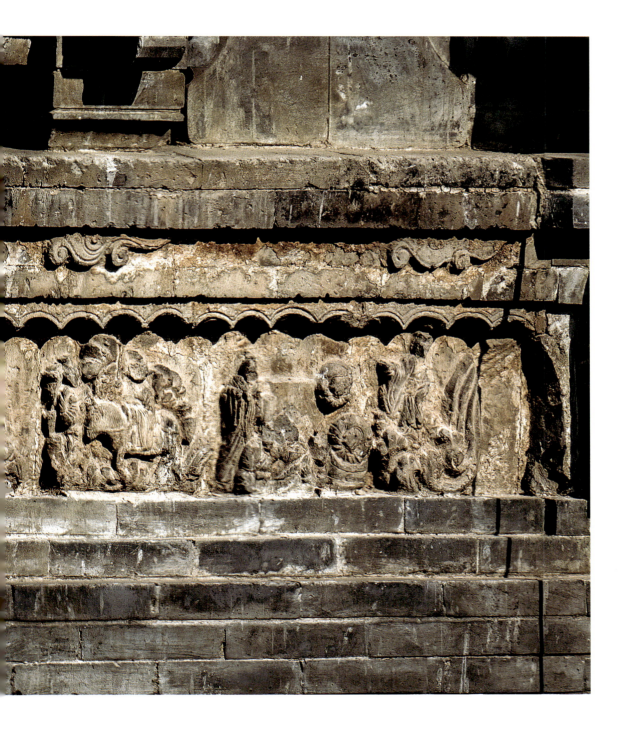

须弥座西面（左）和西南面（右）护法狮子
喀左精严禅寺塔

Guardian lions on the west side (left) and
southwest side (right) of sumeru podium
Pagoda of Jingyan Temple in Kazuo

须弥座西南面护法狮子
喀左精严禅寺塔

Guardian lion on the southwest side of sumeru podium
Pagoda of Jingyan Temple in Kazuo

须弥座西面护法狮子
喀左精严禅寺塔

Guardian lion on the west side of sumeru podium
Pagoda of Jingyan Temple in Kazuo

237

须弥座西北面护法狮子之一
北京房山云居寺北塔
Guardian lion on the northwest
side of sumeru podium (first)
North Pagoda of Yunju Temple in Fangshan,
Beijing

须弥座西南面护法狮子
房山云居寺北塔
Guardian lion on the southwest
side of sumeru podium
North Pagoda of Yunju Temple in Fangshan

239

须弥座西面狮子与舞人之一
房山云居寺北塔

**Dancer and lion on the west side of
sumeru podium (first)**
North Pagoda of Yunju Temple in Fangshan

须弥座西面狮子与舞人之二
房山云居寺北塔

**Dancer and lion on the west side of
sumeru podium (second)**
North Pagoda of Yunju Temple in Fangshan

241

须弥座东北面护法狮子

房山云居寺北塔

Guardian lion on the northeast
side of sumeru podium

North Pagoda of Yunju Temple in Fangshan

须弥座西北面护法狮子之二

房山云居寺北塔

Guardian lion on the northwest
side of sumeru podium (second)

North Pagoda of Yunju Temple in Fangshan

243

245

245

须弥座东北面护法狮子之一
辽宁海城金塔
Guardian lion on the northeast side of sumeru podium (first)
Jin Pagoda in Haicheng, Liaoning

363

须弥座东北面护法狮子之二
海城金塔
Guardian lion on the northeast side
of sumeru podium (second)
Jin Pagoda in Haicheng

须弥座东面护法狮子
海城金塔
Guardian lion on the east side
of sumeru podium
Jin Pagoda in Haicheng

246

248
须弥座南面护法狮子
海城金塔
Guardian lion on the south side
of sumeru podium
Jin Pagoda in Haicheng

249
须弥座北面护法狮子
海城金塔
Guardian lion on the north side
of sumeru podium
Jin Pagoda in Haicheng

248

须弥座东南面护法狮子
海城金塔
**Guardian lion on the southeast side
of sumeru podium**
Jin Pagoda in Haicheng

须弥座东南面护法狮子头部
海城金塔
**The head of guardian lion on the
southeast side of sumeru podium**
Jin Pagoda in Haicheng

250

附　录
名词简释（佛学、历史学、建筑学）*

1（英文条目3）

阿修罗： 古印度神话中的一种鬼神。佛经中常讲到佛说法时，有众多阿修罗齐聚周围。因此佛教石窟艺术中常常出现阿修罗的形象，安置在石窟门的两侧或窟顶上，以示他护卫佛法之意。

2（英文条目13）

宝盖： 佛道或帝王仪仗用的伞盖。盖顶近平，周围有绸边下垂（帷）。上面或缀以铃铛珠玉。塔身佛、菩萨头上的宝盖悬空，略去伞柄结构。

3（英文条目5）

薄伽（薄迦）： 吠陀神话中的阿底提耶众神之一。薄伽一词常表示财富、荣誉、勇敢和美丽。不论人、神，谁有这些优点谁就成为尊者。

4（英文条目38）

椽（chuán）/檐椽/飞檐椽（飞子）： 椽是屋顶结构中架设在檩木上的木条，其功用是承受铺在屋顶的望板和瓦。檐椽是檐下的木椽，承受当檐部分的屋面板及瓦件。飞檐椽是在檐椽前端增设的一段短椽，其功用在于增加屋檐前伸的长度，并使檐端能略微翘起，有助于雨水下流时远射，以保护屋基。

5（英文条目35）

攒尖式顶： 屋顶形式之一。是平面为圆形、方形或其他正多边形的建筑物上的斜坡式屋顶。斜坡较陡，无正脊，数条垂脊攒聚于顶心并呈尖状隆起，故称"攒尖"，上面再覆以宝顶。多用于亭、阁之类，有时亦用于宫殿建筑，如北京的天坛祈年殿等。佛塔的塔顶都是攒尖式顶。

6（英文条目4）

大我/净我/法我/人我/我执： "大我"是在儒学中最先提出，后被逐渐放大的一种观念。如同常说的舍己为人，天下为公。佛教提倡"净我"，即人心中纯净的"我"，要求人剔除心中的妄念与情欲，内心清净。"法我"指支配人和事物的内部主宰者。五蕴（色蕴、受蕴、想蕴、行蕴、识蕴）生灭之法，怖畏生死，妄取涅槃，是为法我见（即虽不信人间俗事，但执念生死伦常，信奉法度，以法为自身主宰）。"人我"指支配人和事物的内部主宰者。"人我"即"我谓主宰"（即简单理解的凡人，舍不下身外之物，以自我为存在的中心和意义）。"我执"指对"我"的执着。认为人不懂缘起无常之理，执着于自身，形成烦恼，造种种业，进而生死轮回，陷入无边苦海。因此"我执"是万恶之本，是一切痛苦和烦恼的总根源。

7（英文条目48）

地宫： 高等级建筑物的地下小室或密室。佛塔的地宫主要是收藏象征佛涅槃的贵重仪物及施主奉献的珍贵物品。

8（英文条目16）

叠涩： 砖或石砌的建筑物，每一层砖、石比下一层砖、石挑出一定的宽度，用以挑出屋檐、平座等的一种砌筑方法。常见于砖塔、石塔、砖石墓室等。

9（英文条目7）

斗栱/铺作/柱头铺作/转角铺作/补间铺作： 斗栱是中国传统建筑的一种结构部件。在立柱与横梁交接处或两条横枋之间，加置一层层探出成弓形的臂式承重结构叫栱，栱与栱之间、栱枋之间垫置的方斗形木块叫斗，合称斗栱。不同位置的斗栱又各有自己具体的名称。铺作是宋式建筑中对斗栱的称呼，此词的由来是指斗栱由层层木料铺叠而成。建筑物柱头上的斗栱称柱头铺作，转角柱上的斗栱称转角铺作，柱头之间的斗栱称补间铺作。

10（英文条目15）

法相： "法"指包括物质和精神在内的一切事物和现象，"相"即形相、相状等。"法相"泛指事物的相状、性质、概念等。

11（英文条目44）

《法苑珠林》： 书名，唐僧道世撰。内容主要是宣传佛家的因果报应，也收有一些民间故事、寓言。

12（英文条目18）

枋： 方体长身的木构件，可有厚薄粗细之分。起水平方向的连接和承重作用。

13（英文条目21）

飞天： 佛教壁画或石刻中空中飞舞的神仙。

14（英文条目8）

佛： （1）指佛教修行达到最高佛果地位的人。义为"觉悟的人"。（2）小乘佛教对释迦牟尼的尊称。大乘佛教则泛指一切觉行圆满者。

15（英文条目34）

覆钵： 原为印度山基大塔、土塔之主体形状，因似扣覆之钵，故名。后辗转变化为中国式塔刹之一部。

16（英文条目11）

浮雕/高浮雕/浅浮雕/透雕： 浮雕指在一块平面的背板上将要塑造的形象雕造出来，使它浮出原来材料的底面，呈凹凸起伏的立体形态，故称浮雕。通常只有一到两面可供观赏。高浮雕指在浮雕作品中起伏甚大，其雕刻凸度可超过雕像周围平面高度的一半，近似于圆雕和半圆雕。浅浮雕是在背景上刻出凸度较小物象的雕刻技法。透雕是一种镂空即通透式雕刻的技法。

17（英文条目22）

浮屠： 或作"浮图""佛图"。梵文Buddhastupa（佛陀窣堵波）音译之讹略，即佛塔。

18（英文条目10）

供养人： 即自愿供养佛、菩萨的人。其形象是当时现实世界的人物，穿着当时的服饰，有明确的社会身份，多为出资建造佛窟的窟主、建造经幢的幢主或施主。对佛、菩萨供养，可求福报（或冥福）。佛塔上的供养人常有较多的人物形象，"供养"的意义也更宽泛。

---

*"名词简释"中文按首字的拼音排序，英文按单词字母排序，序号不对应，特此说明。

19（英文条目25）
勾栏：亦作"钩阑"，宋辽时期及以前对栏杆的称呼。

20（英文条目26）
沟纹砖：在砖的平面（一面）剔出5~10行（或更多）平行的宽约如指、深约1厘米的凹陷纹路，以方便砌垒时与他砖黏合，这种砖称沟纹砖。面有沟纹，是辽砖的特点。

21（英文条目27）
骨朵：古兵器、刑具，棍棒之属。其制由一长棍，顶端套置一锤状物所构成，辽代多用之。

22（英文条目1）
化生童子：化生是一佛教术语。佛教宣扬人有四生："一曰胎生，二曰卵生，三曰湿生，四曰化生。"化生是指无所依托，借业力（佛教中推动生命延续的力量）而出现者。"童子"这一形象则来自犍陀罗美术中"扛花环的童子"，代表着死后的荣光，是新生、再生的象征。莲花蔓草象征丰饶多产。因而在佛教的艺术作品中为了形象地表现由前生修行的业力而化生出的形象，常会在莲花含苞或刚开的莲花中画一些或坐或立的童子，人们称之为化生童子。"金塔"以化生童子荷"人""我"二字，则又增加了哲学意味。

23（英文条目23）
黄閟（bēng）冈下得宝墨：出自宋人楼钥诗《钱清王千里得王大令保母砖刻为赋长句》，记叙的是宋嘉泰二年（1202年），钱清人王畿在黄閟冈下寻得王献之书写的《保母帖》砖刻残碑之事。钱清，今浙江省钱清县。王畿，字千里。王献之，晋代著名书法家，因与其堂弟王珉都曾任中书令一职，二人分别被称为王大令与王小令。《保母帖》为王献之的著名书法作品之一。

24（英文条目28）
减地平钑（sà）：指建筑装饰雕刻的一种形制，即凸起的雕刻面或下凹的底子均为平面。雕刻主体轮廓整齐清晰，有如剪影。

25（英文条目9）
经幢：中国古代的一种宗教石刻。原是建于佛前的丝帛制成的伞盖状物"幢"，周边的幢幡下垂较长，上书佛经经文供人诵读，故名经幢。后来改为石构，柱状，立于佛殿前和路旁，经文改为石刻，称石经幢。是用以宣扬佛法的纪念性建筑物。后来又发展为上下两三层结构，加刻佛像、供养人等内容，幢身下设须弥座，幢顶仿塔顶建筑样式，遂与塔形相近，有的径直称为经塔。佛塔地宫内出土的金银塔大都是经塔。墓葬内有时也立有经幢。

26（英文条目29）
袈裟：佛门高僧做法事时的法服，穿着以示仪式的庄严隆重。

27（英文条目33）
龛：专为供奉佛像、神位而制作的阁子形小型建筑或物品。塔身的佛龛面宽甚窄而进深甚浅，仅容佛像而已。

28（英文条目2）
阑额/普柏枋：阑额是连接檐柱柱头之间的极为厚大的矩形长身木构件，对建筑物柱群的头部起固定作用，从而保持建筑物的稳固。清代名额枋。普柏枋是垫在阑额和柱头之上以承托斗栱的长板状木构件。清代名平板枋。

29（英文条目17）
灵塔：原指佛涅槃后收藏佛舍利与佛骨之塔，后为纪念佛教八大圣地，亦以这些圣地为这些灵塔命名。八大圣地是有关佛祖释迦牟尼从出生到涅槃八个阶段的重要场所。

第一座塔：净饭王宫生处塔。释迦降生。释迦牟尼29岁时，痛感人生生老病死的各种烦恼，所以他舍弃了王族生活，出家修道。

第二座塔：菩提树下成佛塔。降魔成道，释迦牟尼在菩提树下，豁然开悟如人醒来的彻悟境界，由此成道。

第三座塔：鹿野苑中法轮塔。初转法轮，纪念释迦牟尼在鹿野苑中开始传教。

第四座塔：给（jǐ）孤独园名称塔。释迦牟尼曾在给孤独园说法示现大神通。

第五座塔：曲女城边宝阶塔。所讲故事为佛上忉（dāo）利天为母说法，又从三十三天降凡。

第六座塔：耆阇崛山般（bō）若（rě）塔。为佛化度处，即王舍城，佛常居此弘扬佛法的地方。

第七座塔：庵罗卫林维摩塔。讲的是维摩居士卧于病榻之上，文殊菩萨奉释迦命前去问疾，讨论佛法的故事。

最后一座塔：娑罗林中圆寂塔。娑罗树也叫八叶树，被称为圣树。释迦牟尼80岁时，在娑罗双树下涅槃（去世）。

30（英文条目24）
鎏金：用金质材料涂附在其他金属器物上以作为装饰的一种方法。

31（英文条目14）
密宗：中国佛教派别之一。源出于古印度佛教中的密教。唐开元初，善无畏、金刚智、不空三人先后来华翻译传播佛教经典，形成宗派。以《大日经》和《金刚经》为依据，把大乘佛教的烦琐理论运用在简化通俗的颂咒祈祷方面。认为口诵真言（语密）、手结契印（身密）、心作观想（意密）三密同时相应，可以即身成佛。公元8~11世纪，印度密教传入中国。

32（英文条目36）
平座：建筑物上的一层平台。或用柱、枋、斗栱等架起或不用。砖塔上的平座也是砖仿木结构，上承塔身。

33（英文条目6）
菩萨：菩萨是梵文Bodhisattva音译菩提萨埵（duǒ）的略称，意指通过修持以求无上菩提（觉悟），可利益众生，于未来能成就佛果的修行者。

34（英文条目12）
石窟寺：依山崖开凿建造的佛教寺庙建筑。起源于印度，约在东汉以后随佛教经西域传入中国内地。窟内多以雕塑或壁画描绘佛像和佛经故事等，是研究佛教史和古代社会生活的重要资料。

35（英文条目41）
收分：逐步收缩减少的意思，是古建筑及其他物事使墙厚、柱径、碑宽下大上小，墙面、柱面微向内倾的做法。

36（英文条目31）
手印/智拳印：手印原意为手势，即以手臂尤其是手指做成的具有特定意义的各种姿势、形态。在古印度，手印广泛应用于计算、舞蹈、记诵、雕塑、宗教仪式等领域。在密教中，手印表示佛、菩萨、诸天尊基于自己对佛教义理的解悟而做出的自誓。智拳印也称最上菩提印，是毗卢遮那佛专用的手印，代表他的智慧。手印形态为两拳放在胸前，左拳伸出食指，以右拳握之。

37（英文条目39）
舍利/舍利塔：舍利是梵文Sarira的音译，意为佛骨。相传为释迦牟尼遗体火化后出现的不同颜色的粒状结晶体。高僧遗体火化有时也出现过舍利。供奉佛舍利的塔，称舍利塔。

38（英文条目43）
素平/线雕/阴文：素平是仅有线雕装饰纹样的平面。线雕是用阴线（凹陷）或阳线（凸起）作为主要造型手段的雕刻技法。阴文是玺印或其他器物上刻铸成凹下的文字或花纹。

39（英文条目19）

塔刹/相轮： 塔刹是塔身顶端直立的细而长的杆状物。刹是梵语"刹多罗"的简称，原意是田土、国土，也表示佛国、佛寺，具有佛教的象征意义。相轮为塔刹的主要部分。是由刹杆串联之若干圆盘（轮）状物，其数量多与塔的层数相应。

40（英文条目46）

剔地起突： 建筑雕刻中最复杂的一种，类似于高浮雕或半圆雕。特点是雕刻母题突出底面较高，"地"即底子被剔除而层层凹下，起伏大，层次多。雕刻的最高点不在同一平面上，雕刻的各部位可以互相重叠交错。

41（英文条目30）

天王/力士： 天王本是在印度神话中惩恶护善的人物，佛教称之为"天"（梵名Deva），是护持佛法的天神。力士指有大力气的人。

42（英文条目47）

箅瓦/板瓦/滴水： "箅"字又作"筒"，一种半圆筒形的瓦。常与板瓦（一种横断面小于半圆的、弧片状的瓦）上下扣合使用，构成一条条的瓦垄，是屋面的主要防水构件。滴水是安放在屋面板瓦垄沟最下端出檐处的一种排水构件。雨水自此流下，故名滴水。

43（英文条目37）

槫（tuán）： 即桁（héng）或檩（lǐn），圆体长身的厚重木构件。横架在屋架与山墙上，支持椽子或屋面板起承重作用。

44（英文条目20）

五方佛： 即五方如来。源自密宗金刚界思想，东西南北中五方，各有一佛主持。位于中央的为大日如来，又作毗卢遮（舍）那，意译为"光明遍照"。另有东方阿閦（chù）佛（不动如来佛）、西方阿弥陀佛（无量寿佛）、南方宝生佛（宝相佛）、北方不空成就佛（微妙声佛），他们不是独立存在。五方如来是对"佛"的概念的抽象表述。

45（英文条目42）

须弥座/壶门/瘿项柱： 须弥座是古建筑中一种形式的台座。常用于神龛、佛坛、台、塔、幢以及其他等级较尊贵的建筑物。侧面呈"工"字状，中有束腰。束腰与上下台面的连接处有时为坡形，称上、下枭。在塔的须弥座束腰处往往做出一些壶门，这是一种小而有一定深度的门洞，门内外又有雕饰。瘿项柱是隔开壶门与壶门之间的短柱，柱甚矮而中腹膨出，状如人之瘿项，故名。柱上亦有雕饰。

46（英文条目49）

压地隐起： 古建筑雕刻形式之一，是一种浅浮雕。装饰面无论是平面还是弧面，雕刻各部位的高点，几乎都在同一水平上。当雕饰面有边框时，各部位高点不超出边框的高度。雕刻各部位互相重叠穿插，有一定的深度感。

47（英文条目50）

遥辇痕德堇： 契丹遥辇氏部族首领。遥辇为姓氏，痕德堇为名。

48（英文条目45）

药师佛/药师七佛： 药师佛为梵文Bhaisajya-guru（鞞杀社窭鲁）的意译，具名为"药师琉璃光如来"，又称"大医王佛"。大乘佛教的佛名。《药师经》说，此佛曾发十二心愿，医治众生病苦，消灾延寿。在一些寺院的大殿中，其塑像常与释迦、弥陀二佛并坐，为"横三世佛"。其中药师佛主管东方净琉璃世界，为东方佛；释迦、弥陀为中土与西方佛。药师七佛即七尊药师如来：善名称吉祥王如来、宝月智严光音自在王如来、金色宝光妙行成就如来、无忧最胜吉祥如来、法海雷音如来、法海胜慧游戏神通如来、药师琉璃光如来。

49（英文条目51）

耶律阿保机： 辽朝开国皇帝，阿保机为名，耶律为姓氏。公元907~926年在位。

50（英文条目52）

夷离堇： 辽代官名，契丹最高军事首领。

51（英文条目32）

璎珞： 以珠玉等穿缀的颈项串饰。在绘画或雕塑中，为神佛、菩萨天女等所佩用，后来亦成为仕女之佩饰。

52（英文条目40）

圆雕/半圆雕： 圆雕是不附着在任何背景上，各面均雕出的完全立体的雕像，可从任何角度观赏。半圆雕则是不脱离原石、原木、原器物、原衬景与原烘托纹饰的雕塑，即仅有一半是圆雕，只能从正面和侧面观赏。

53（英文条目53）

《卓歇图》： "卓"是立的意思，"卓歇"就是支起帐篷休息。此图是一幅生动地描绘契丹族部落猎后休息娱乐情景的绘画长卷，作者为五代时期的画家胡瓌。

APPENDIX

Definition (Buddhism/Historiography/Architecture)

1

Anupadaka (self-created) Child: Anupadaka is a Buddhist term. According to Buddhist doctrine, there are four forms of people's birth: jarayuja (viviparous), andaja (oviparous), samsvedaja (moisture, or water born) and anupadaka (self-produced or self-created). The term 'anupadaka' refers to certain celestial beings that are not emanated by higher entities, but are 'self-born' from the divine essence. The image of kid comes from the Gandhara art, a boy wearing a wreath, representing the glory after death, is a symbol of rebirth and regeneration. The lotus with creeping weeds signifies fertility. Images of the boy standing on blooming or budding lotus are frequently in the art of Buddhism. Such motif is for representing the rung of anupakada by karma of previous life's practicing.

2

Architrave/ Pubai Tie-beam: Architrave is a thick rectangular timber component to connect the capitals of eaves' columns. It can fix the column heads group to keep the building stable, and is called 'E Fang' in the Qing Dynasty. 'Pubai tie-beam' is a long slab timber on the architraves and column capitals to retain the bracket set. It is called 'Pingban Fang' in the Qing Dynasty.

3

Asura: Asura is a demigod in ancient Indian mythology. Buddhist scripture often says that there were a lot of Asuras in the vicinity when Buddha taught the Dharma. Therefore, the art of Buddhist grottoes often appear the image of Asuras, placed in the grotto on both sides of the door or cave, and where they are available to protect the Buddhism.

4

Atman (self): 'Great Self' means that one has a mind which sacrifices oneself for others and the world. 'Atman' in Buddhism is the concept of self, or ego, which refers to a person's 'true self', 'thinker of thoughts', 'feeler of sensations'. In Buddhism, 'Atman (self)', as a sentient being's person and personality, is generated in harmony from five skandhas (form, sensations, perceptions, mental activity and consciousness). 'Purifying one's mind' or 'one's practicing' is to get rid of delusions and passions in the heart, and to obtain inner purity. 'Eso me atta(atman)' or 'this is mine', means that a physical person can not abandon the things outside body, and take self as the center and meaning of existence. 'Atman-graha' or 'I am my own master', means vasana (habitual tendencies) of self-attachment. Buddhism

is believed that sentient beings do not understand the reason of impermanence and persist in themselves. So let themselves get into trouble and experience all kinds of difficulties. At the last, they pass into the cycle of birth and death. The pain will never ends. Thus 'atman-graha' is the root of all evil, pain and trouble.

5

Bhaga: One of the Adityas in the vedic myth. The word often means wealth, honor, courage and beauty. No matter man or god, who has these virtues will become a respected person.

6

Bodhisattva: Pu Sa(菩萨) in Chinese is the abbreviation of the transliteration from Sanskrit 'Bodhisattva', which means a practitioner who has generated bodhicitta, a spontaneous wish and compassionate mind to attain Buddhahood for the benefit of all sentient beings.

7

Bracket Set/ Bracket Set on Column/Bracket Set on Corner/Bracket Sets between Columns: Bracket set is one of structural components of Chinese traditional architectures. At the junction of the vertical column and the horizontal beams, or two horizontal tie-beams, is the arched bracket called the 'Gong'. The square wooden block that between two arches or between the Gong and Fang tie-beam is called the 'Dou'. Gong and Dou make up the 'bracket set'. Bracket sets in different locations also have their own specific names. Bracket set is called 'Pu Zuo' in the Song Dynasty. The bracket set on column head of the building is called 'bracket set on column', on the corner column called 'bracket set on corner', and between two column-heads called 'bracket sets between columns'.

8

Buddha: (1) It refers to the person who achieves the highest level by Buddhist practices. It means the 'awakened one'. (2) Hinayana Buddhism refers to the Sakyamuni's name. Mahayana Buddhism refers to all people who have known and done everything.

9

Buddhist Stone Pillar: It is a religious stone carving in ancient China. It was originally shaped like a silk canopy in the front of Buddha and written the sutra for people to read. Then it took on the appearance of a high stone column erected in front of the temple or beside the road. It was a commemorative architecture to promote the dharma, and later developed into two or three layers

of structure with the Buddhas and donators carved on surfaces. With a sumeru podium built as a base, its top likes the top of tower, so it is similar to the tower and some directly called sutra tower. Most of the gold and silver pagoda unearthed in the underground palace of the pagoda are shaped of sutra towers. Some tombs also have Buddhist stone pillars erected inside.

10

Buddhist Donators: Donators are those who provide the Buddha or Bodhisattva of their own accord. Their images are real figures wearing the clothes at the time and have clear social identities. Most of them are always the owners or benefactors of constructing the grottoes and Buddhist stone pillar. They provide the Buddha and Bodhisattva to ask for blessed rewards. There are many kinds of donator's visualizations on the pagodas, so the meaning of donation or devotion is also broader.

11

Carved Relief/ High Relief/ Bas Relief/ Open-work: Carved relief refers to carve the image that the sculpted material has been raised above the background plane. It is available for viewing only by one or two sides. High relief has great fluctuation, where more than 50% of the depth is shown. It is similar to the round relief and half round relief. Bas relief (low relief) is a engraving technique, where the plane is only very slightly lower than the sculpted elements. Open-work is a technique that produces decoration by creating holes, piercing, or gaps that go right through a solid material.

12

Cave Temple: The Buddhist temple carved out of stones on the hill cliff is originated from India. It was introduced into China's interior through the Western Region after the Eastern Han Dynasty. Inside cave are sculptures or murals, which depict the Buddha statues and Buddhist stories. They are taken as important materials for the study of Buddhist history and ancient social life.

13

Ceremonial Canopy: It is a canopy used by Buddha or emperors. The top of canopy is nearly flat and has silks hanging around. The canopy on Buddha's and Bodhisattva's head is suspended and omitted the structure of a umbrella handle.

14

Chinese Esoteric Buddhism: *Tantrism, as one of major branches in Chinese Buddhism, denotes the esoteric tradition of Buddhism that developed in India. During the Kaiyuan Era of the Tang, three great masters Subhakarasimha, Vajrabodhi and Amoghavajra came to China. They translated key Buddhist texts and founded the 'Zhenyan' ('true word', 'mantra') tradition. This tradition focused on tantras like the Mahavairocana Tantra and the Vajracchedika Sutra (Diamond Sutra), applied the tedious theories of Mahayana Buddhism to the use of mantras. Buddhist esotericism is centered on what is known as 'the three mysteries' or 'secrets': mudra (the Secret Body), mantras (the Secret Speech), and samadhi (the Secret Mind). Through these unique methods, a practitioner is to be provided with a faster path to Buddhahood. Indian Tantrism was introduced to China in the 8th–11th century.*

15

Dharma Lakasana: *'Dharma' means everything and phenomena, including material and spirit. 'Lakasana' means characteristic, trait and feature. 'Dharma lakasana' refers to the aspects of things, including nature, concept and characteristic.*

16

Die Se (corbelling): *The technique of corbelling usually employed in the building of stone and brick, where rows of stones or bricks deeply keyed inside a wall support a projecting parapet, eaves and platform. It is common in brick or stone towers as well as brick tombs.*

17

Divine Pagodas: *They were originally built to collect Buddhist relics and Buddha bones after the nirvana of Buddha. Then in order to commemorate eight pilgrimage sites of Buddhism, these pagodas are named after these holy places. The sites are important places for Sakyamuni from birth to nirvana.*
The first is called 'the Maya Devi Temple', a site traditionally considered the birthplace of Gautama Buddha. When he was 29 years old, he felt all the worries of life. So he abandoned the royal family to cultivate.
The second is called 'the Mahabodhi Temple'. It is the place where Gautama attained the Enlightenment under what became known as the bodhi tree, so he became a Buddha.
The third is called 'the Dharmarajika Stupa' in Sarnath, where Gautama Buddha first taught the Dharma.
The fourth is called 'the Jetavana Vihara', where Sakyamuni gave the majority of his teachings and discourses, and showed the magical power.
The fifth is called 'the Visahari Devi Temple' in Sankissa. The Buddha ascended to the Trayastrimsa (belonging to the thirty-three Heaven) for his mother. Then he taught her abhidharma.
The sixth is called 'the Monastery Venuvana Vihara'. It is the place where Buddha superseded Rajgir (historically known as Girivraj) and often resided to propagate Dharma.
The seventh is called 'the Vimalakirti Stupa' in Vaisali. A story was told there that Manjushri Boddhisattva visited to inquire about a disease and discuss the Buddhist Dharma when the layman Vimalakirti was ill.
The last is called 'the Mahaparunirvana Temple'. Shorea robusta, the sol tree is named the 'eight leaf tree' and is
called the sacred tree. When the Buddha was 80 years old, he was lying between a pair of sal trees when he died.

18

Fang (tie-beam): *Fang are long and cuboid structural elements and differ in thickness. They are functioned as horizontal connection and load-bearing.*

19

Finial/ Sorin (alternate rings): *The finial (Ta Cha) is a long and thin, vertical shaft that tops a pagoda. Cha (刹) is the abbreviation from the transliteration of the Sanskrit 'ksetra'. Ksetra refers to 'field' or 'tract of land', and in a broader sense represents 'Buddhist temple' or 'residence of Buddha'. The sorin, functioned as a main part of the finial, is composed of several alternate rings or wheels, and its number is the same as that of pagoda's levels.*

20

Five Tathagatas: *'The Five Tathagatas' or 'the Five Great Buddhas', are a development of the Buddhist Tantras, and first appear in the Vajrasekhara Sutra. They are aspects of the dharmakaya or dharma-body, and related to the five directions of east, west, south, north and center. Vairocan, or Vajradhra is at the center, while Amitabha in the West, Akshobhya in the East, Ratnasambhava in the South, Amoghasiddhi in the North. They are not an independent existence but an abstract expression of the concept of Buddha.*

21

Flying Asparas: *Flying Asparas is a type of celestial being spirit who prominently figures in the Buddhist murals and stone carvings.*

22

Futu: *'Futu' or 'Fotu' is an incorrect abbreviation from the transliteration of the Sanskrit 'Buddhastupa'.*

23

Getting the Precious Calligraphy beneath the Huangbeng Mound: *This sentence is from 'The Poem for Qianqing's Wang Qianli Getting the Bao Mu Brick with Wang Daling's Calligraphy', by Lou Yao of the Song. It tells the story of Qianqing's Wang Ji (courtesy name is Qianli) who found a brick engraved with Wang Xianzhi's Bao Mu Calligraphy beneath the Huangbeng Mound in 1202. Qianqing is now Qianqing county in Zhejiang Province. Wang Xianzhi is a famous calligrapher in the Jin Dynasty. He and his brother Wang Min both held the position 'Zhongshu Ling' (head of the secretariat), so they were called 'Wang Daling' and 'Wang Xiaoling' (meaning the older one and the younger). 'The Bao Mu Calligraphy' is one of the most famous calligraphies by Wang Xianzhi.*

24

Gilding: *It is a decorative technique for a very thin coating of gold to solid other metal surfaces.*

25

Gou-lan (balustrade): *'Gou-lan' is a Chinese architectural term refers to a balustrade during and*
before the Song and Liao Dynasties.

26

Groove Bricks: *In order to facilitate the adhesion between the bricks, the plane (one side) of the brick is carved 5 to 10 lines (or more) what the width likes a-finger's and the depth is about 1 cm. It is named groove brick. The surface with groove is the characteristic of bricks in the Liao.*

27

Gu Duo: *It is a kind of cudgel ancient weapon or an instrument of torture. It is made of a long stick with a hammer-shaped head. Gu Duo often used in the Liao Dynasty.*

28

Jian-di Ping-sa (flat carving): *It is a form of architectural decorative carvings. Whatever the raised carving surface or background plane are both flat, and the outline of the carved motif is as regular and clear as a silhouette.*

29

Kasaya: *It is the dress for the Buddhist monks when they are doing religious rites, in order to show the solemnity.*

30

King of Heaven/ Li Shi (strong man): *The King of Heaven is originally a figure who punishes the evil and protects the good in Hindu mythology. 'Heaven' in Buddhism means 'Deva' (Sanskrit) and King of Heaven is a supernatural warrior for guarding Buddhism. 'Li Shi' means a man of great strength.*

31

Mudra/ Vajra Mudra: *A mudra is a symbolic or ritual gesture or pose in Buddhism. While some mudras involve the arms, most are performed with hands and fingers. In ancient India, mudras have been employed in many fields such as dance, calculation, recitation, sculpture and religious ceremonies. In tantra, mudras represent the enlightenment and oath of the Buddha, Bodhisattva and deities based on the Buddhist doctrine. Vajra Mudra or Jnana Mudra or Bodhasri Mudra, only used by Vairocan, indicates supreme knowledge or wisdom. It is made by forming two punch in the chest, while a fist with left hand but extending the first finger, and then grasping that finger with the right hand.*

32

Necklace of Jade and Pearls: *It is a necklace with beads and jade and so on. In painting or sculpture, it is for gods and goddess to wear. Then, it also became the female adorning.*

33

Niche: *A niche is a small lattice-shaped structure that consecrate the statue of Buddha or other spirits. The niche inside pagodas is very narrow and deep. It is only enough for a Buddha statue.*

34

Overturned Bowl: This was originally shaped as the mound-like or hemispherical pagoda in India. It's named because it looks like an overturned bowl. Then it evolved into a key component of the Chinese pagoda finial.

35

Pavilion Roof: It is one of the roof forms on the buildings with a plane of circular, square or other regular polygon. It's a relatively steep slope without main ridge and gathered many ridges in the center of the roof and showed an uplift sharp, so it's called 'pavilion roof'. It is mainly used for kiosks, lofts and some palaces, such as the Hall of Prayer for Good Harvest (the Temple of Heaven in Beijing). The Buddhist pagodas are all constructed with pavilion roofs.

36

Ping Zuo (platform): It is referred to a platform projecting from the wall of a building, supported by columns and brackets or not. The Ping Zuo of brick towers is also brick imitation of wood structure, and is commonly functioned for supporting the body.

37

Purlin (Tuan): It is also called 'Heng' or 'Lin' in Chinese architectural terms. As thick, round wood frame parts, purlins are mainly laid across the roof trusses and gables to support the rafters at roof decks.

38

Rafter/ Eaves Rafter/ Flying Rafter: A rafter is one of a series of sloped battens attached to the purlins in the roof structure in order to support the roof deck and tiles. Eaves rafters are designed under the eaves, in order to support the wall cladding and tiles. The flying rafters are short ones that extend just to the edge of the eave roofs, in order to increase the length of eaves and make them curve upward. This curvature is helpful for protecting the roof when the rain came down.

39

Sarira (She Li)/ Sarira Stupa: 'She Li' is a transliteration of the Sanskrit 'Sarira', which means the Buddha bone. It usually refers to colorful crystal-like bead-shaped objects that are purportedly found among the cremated ashes of Sakyamuni. Relics of Buddhist spiritual masters after cremation are also discovered. A stupa is a shire containing the Buddhist relics.

40

Sculpture in the Round/ Mid-relief Sculpture: Sculpture in the round is a type of sculpture in which the figures are presented in complete three-dimensional form and are not attached to a flat background. So people can watch it from any angle. Mid-relief sculpture is only where up to half of the subjects project from the ground, in addition none of the sculpted elements are undercut from the original stone, logs, artifacts and decorative backgrounds. And the images can be only seen from the front and the sides.

41

Shou Fen (entasis): In architecture and other things, entasis is the application of a convex curve to a surface for aesthetic purpose. Its best-known use is in certain orders of classical columns, walls and tablets that curve slightly as their diameter and width decreased from the bottom upward, or their thickness shrunken from the outer inward.

42

Sumeru Podium/ Kun Men/ Ying-xiang Pillar: The sumeru podium refers to a base in ancient architecture. It is often used for niche, altar, tower, Buddhist stone pillar and other high-leveled architectures. It is shaped like a Chinese character ' 工 ' for viewing from side aspect. Some of podiums are designed with the upper and lower cymas on both ends of the narrow middle part. 'Kun Men' means a small and deep niche-shaped door, and is usually carved in the middle part of a sumeru podium. The Ying-xiang Pillar, as one of architectural members characterized by its shortness and median eminence, is designed for dividing the two 'kun men' doors from each other. 'Ying-xiang' is a Chinese term that means 'a goiter neck', so the pillar of such kind is named after it. And it always has patterns decorated on the pillar.

43

Su Ping/ Line Engraving/ Intaglio: Su Ping is a Chinese term for surfaces of a work engraved only with lines. Line engraving is a relief technique, with the images made by concaved or convex lines. Intaglio means carving in intaglio, is a technique employed in making seals or others, with the characters and patterns cut into the flat background of a hard substance.

44

A Forest of Pearls in the Garden of Dharma: This is a book written by Sangha Daoshi in the Tang Dynasty. The book mainly promotes Buddhist karma, also has some folk stories and fables.

45

The Medicine Buddha/ Seven Past Buddhas of Medicine: The Medicine Buddha is a free translation of the Sanskrit 'Bhaisajyaguru'. It is formally named 'the Medicine Master and King of Lapis Lazuli Light', or also called 'the Great Medicine King Buddha' in Mahayana Buddhism. According to the *Medicine Buddha Sutra*, this Buddha made twelve great vows to eliminate all the suffering and afflictions of sentient beings, and to prolong their lives. In the main halls of the Buddhist temples, he is typically depicted seating with Sakyamuni and Amitabha side by side, and thus they are called 'three Buddhas of horizontal existence'. Among them, Bhaisajyaguru considered to be the Buddha of the eastern pure land of Vaiduryanirbhasa, while Sakyamuni and Amitabha to be Guardians of the central and the west. The Seven Medicine Buddhas or the Seven Past Buddhas of Medicine are named separately as follows: Tathagatha Glorious Renown of Excellent Signs; Tathagatha King of Melodious Sound; Tathagatha Stainless Excellent Gold; Tathagatha Supreme Glory Free From Sorrow; Tathagatha Melodious Ocean of Dharma Proclaimed; Tathagatha King of Clear Knowing; Tathagatha Medicine Guru Lapis Lazuli King.

46

Ti-di Qi-tu (elevated relief): It is similar to what is commonly referred to high relief or semi-circle carving, which is the most complex of architectural carvings. The characteristic is that the carved motif is much higher than the surface of stone and the bottom layer is concave, and it has a lot of ups and downs. The highest point of the carving is not on the same plane and the various parts can be overlapped.

47

Tubular Tile/ Plate Tile/ Drip Tile: The tubular tiles are used as roof tiles, and each is semicircular in section. They normally with the plate tiles (each less than semicircle and arched in section), fastened up and down, are laid in regular overlapping rows. Both of them are functioned as the main waterproof for protecting the roof surfaces. The drip tile is one of drainage members projecting under the end of tile furrows. The rainwater falls from here, hence the 'drip' is named.

48

Underground Palace: It is an underground cell or chamber of a high-leveled building. The underground palace of the pagodas are mainly used to collect the precious ritual objects that symbolize Buddha's nirvana or valuable things donated by the donators.

49

Ya-di Yin-qi (low relief): As a kind of bas relief or low relief, it is one of sculpted methods in the ancient architecture. Whether the decorative surface is plane or arched, the carved parts are almost on the same level. When the carved image has a border, the highest point of sculpted material is no more than that of the frame. The carved elements are overlapped and interlocked each other which create the illusion of depth for sculpture.

50

Yao-nian Hen-de-jin: The tribal leader of Yao-nian of the Khitan. Yao-nian is last name, Hen-de-jin is first name.

51

Yelv A-bao-ji: The name of the founding emperor of the Liao Dynasty. Yelv is last name, and A-bao-ji is first name.

52

Yi-Li-Jin: This is an official's name of the Liao Dynasty and the supreme military chief of Khitan.

53

Zhuoxie Painting: 'Zhuo' means 'erect'. 'Zhuoxie' means 'set up a tent to rest'. This painting by Hu Gui in the Five Dynasties is a long scroll that vividly depicts the Khitan's recreation after hunting.

本书所收辽代砖塔资料简表
Brief List of Liao Dynasty Brick Pagodas in this Book

图像 PICTURE	名称 NAME	所在地 LOCATION	年代 ERA	塔高 HIGHT	建筑形式 ARCHITECTURAL FORM	备注 REMARKS
	朝阳北塔 North Pagoda in Chaoyang	辽宁朝阳 Chaoyang, Liaoning	辽重熙十三年（1044） In the 13th year of the Chongxi era, Liao Dynasty(AD 1044)	42.6 米 42.6m	方形十三级密檐式，单层塔身，单层须弥座 Squre structure with thirteen levels and multi eaves, single layered pagoda body, single layered sumeru podium.	原建于隋唐时期，辽代补修重建，1988 年重修，发现天宫、地宫、塔心室及两颗"释迦牟尼真身舍利" It was built in the Sui and Tang dynasty and rebuilt in the Liao dynasty. When the pagoda was repaired in 1988, there the heavenly palace, underground palace and two sariras of Sakyamuni were found.
	庆州白塔 Qingzhou White Pagoda	内蒙古巴林右旗 Balin Youqi, Inner Mongolia	辽重熙十八年（1049） In the 18th year of the Chongxi era, Liao Dynasty(AD 1049)	49.48 米 49.48m	八角七级楼阁式，七层塔身，单层须弥座 Octagonal seven-leveled pavilion-like structure, seven layered pagoda body, single layered sumeru podium.	
	广济寺塔 Pagoda of Guangji Temple	辽宁锦州 Jinzhou, Liaoning	辽清宁三年（1057） In the 3rd yaer of Qingning era, Liao Dynasty(AD 1057)	原高 63 米，存高 57 米 The original height 63m, now 57m	八角十三级密檐式，单层塔身，双层须弥座 Octagonal structure with thirteen levels and multi eaves, single layered pagoda body, two layered sumeru podium.	1996 年大修 It underwent major repairs in 1996.
	蓟县白塔 White Pagoda in Jixian	天津蓟县（蓟州区） Jixian, Tianjin	辽清宁四年（1058） In the 4th yaer of Qingning era, Liao Dynasty(AD 1058)	30.6 米 30.6m	八角覆钵式，单层塔身，单层须弥座 Octagonal overturned bowl-like structure, single layered pagoda body, single layered sumeru podium.	1976 年唐山地震，塔身受损严重，1983 年大修 The pagoda body was severely damaged when the earthquake occurred at Tangshan in 1976. Then, it was repaired in 1983.

图像 PICTURE	名称 NAME	所在地 LOCATION	年代 ERA	塔高 HIGHT	建筑形式 ARCHITECTURAL FORM	备注 REMARKS
	精严禅寺塔 Pagoda of Jingyan Temple	辽宁喀左 Kazuo, Liaoning	辽咸雍五年 （1069） In the 5th year of the Xianyong era, Liao Dynasty(AD 1069)	残高 34.1 米，修缮后约 40 米 The residual height was 34.1m, about 40m after repaired.	八角九级密檐式，双层塔身，原应为三层须弥座，中下层座残损。为楼阁式与密檐式结合典型实例 Octagonal structure with nine levels and multi eaves, two layered pagoda body, three layered sumeru podium, but middle and lower layered sumeru podium were damaged. It is a typical example of the combination of pavilion style and multi-eaves.	清乾隆四十五年（1780）重修，并新建高台 In the 45th year of Qianlong's reign (AD 1780), it was repaired and a new high platform has been built.
	大明塔 Daming Pagoda	内蒙古宁城 Ningcheng, Inner Mongolia	约辽道宗时期 About the Period of Liao Daozong	74 米 74m	八角十三级密檐式，单层塔身，单层须弥座 Octagonal structure with thirteen levels and multi eaves, single layered pagoda body, single layered sumeru podium.	1976 年唐山地震，塔顶倾斜，风铎、铜镜大部坠落，1981 年全面修缮 Because of the earthquake occurred at Tangshan in 1976, the pagoda top tilted and most wind-bells and bronze mirrors fell. Then, the pagoda was repaired in 1981.
	华严经塔 Huayanjing Pagoda	内蒙古呼和浩特 Hohhot, Inner Mongolia	约辽道宗时期 About the Period of Liao Daozong	45.18 米 45.18m	八角七级楼阁式，七层塔身，双层须弥座 Octagonal seven-leveled pavilion-like structure, seven layered pagoda body, two layered sumeru podium.	20 世纪 80 年代重修 It was repaired in the 1980s.
	凤凰山云接寺塔 Pagoda of Yunjie Temple at Fenghuang Mountain	辽宁朝阳 Chaoyang, Liaoning	约辽道宗时期 About the Period of Liao Daozong	37 米 37m	方形十三级密檐式，单层塔身，单层须弥座 Squre structure with thirteen levels and multi eaves, single layered pagoda body, single layered sumeru podium.	此塔形制与朝阳北塔相似，建造年代待考 The shape and structure of this pagoda is similar to the North Pagoda at Chaoyang, but its time of construction needs to be proved.
	云居寺北塔 North Pagoda of Yunju Temple	北京房山 Fangshan, Beijing	约辽道宗时期 About the Period of Liao Daozong	30.4 米 30.4m	覆钵楼阁组合式，双层塔身，双层须弥座 Combination of overturned bowl and pavilion-like structure, two layered pagoda body, two layered sumeru podium.	

图像 PICTURE	名称 NAME	所在地 LOCATION	年代 ERA	塔高 HIGHT	建筑形式 ARCHITECTURAL FORM	备注 REMARKS
	凤凰山大宝塔 Dabao Pagoda at Fenghuang Mountain	辽宁朝阳 Chaoyang, Liaoning	约辽道宗时期 About the Period of Liao Daozong	17 米 （无塔刹） 17m (no pagoda finial)	方形十三级密檐式，单层塔身，单层须弥座 Squre structure with thirteen levels and multi eaves, single layered pagoda body, single layered sumeru podium.	刹顶坍毁无存 The pagoda finial collapsed without any survivors.
	辽上京南塔 South Pagoda at Liao Upper Capital	内蒙古巴林左旗 Balin Zuoqi, Inner Mongolia	约辽道宗或天祚帝时期 About the Period of Liao Daozong or Tianzuo Emperor	25 米 25m	八角七级密檐式，单层塔身，须弥座损坏不存 Octagonal structure with seven levels and multi eaves, single layered pagoda body, sumeru podium was damaged.	塔身除宝盖、灵塔尚较完整外，众多雕塑损毁 Apart from ceremonial canopies and divine pagodas, many statues were damaged.
	天宁寺塔 Pagoda of Tianning Temple	北京 Beijing	辽天庆十年（1120） In the 10th year of Tianqing era, Liao Dynasty(AD 1120)	57.8 米 57.8m	八角十三级密檐式，单层塔身，双层须弥座 Octagonal structure with thirteen levels and multi eaves, single layered pagoda body, two layered sumeru podium.	1976 年唐山地震，全塔受损，塔刹坍塌，刹杆孑立，部分砖雕损毁 The pagoda was damaged, the pagoda final was collapsed except the rod of it, and some brick carvings were destroyed in the 1976 Tangshan Earthquake.
	八棱观塔 Balengguan Pagoda	辽宁朝阳 Chaoyang, Liaoning	约辽天祚帝末年 About the end of Liao Tianzuo Emperor	34.4 米 34.4m	八角十三级密檐式，单层塔身，三层须弥座 Octagonal structure with thirteen levels and multi eaves, single layered pagoda body, three layered sumeru podium.	20 世纪 90 年代，盗塔者为掘地宫炸塔，塔身及须弥座损毁严重，无塔刹，疑为辽末未完工程 In the 1990s, robbers bombed the pagoda to dig underground palace. The body, sumeru podium were badly damaged. The Pagoda finial was not found, It could be an unfinished project during the end of Liao dynasty.
	辽阳白塔 White Pagoda in Liaoyang	辽宁辽阳 Liaoyang, Liaoning	约辽代晚期 About the Late Liao Dynasty	71 米 71m	八角十三级密檐式，单层塔身，双层须弥座 Octagonal structure with thirteen levels and multi eaves, single layered pagoda body, two layered sumeru podium.	1990 年维修，发现金元时期维修所留文字《重修辽阳城西广佑寺宝塔记》 *The Record on Rebuilding the Pagoda of Guangyou Temple at Liaoyang*, the text left in Jin and Yuan period, when the pagoda was repaired in 1990.

图像 PICTURE	名称 NAME	所在地 LOCATION	年代 ERA	塔高 HIGHT	建筑形式 ARCHITECTURAL FORM	备注 REMARKS
	金塔 Jin Pagoda	辽宁海城 Haicheng, Liaoning	约辽代晚期 About the Late Liao Dynasty	31.5 米 31.5m	八角十三级密檐式，单层塔身，双层须弥座 Octagonal structure with thirteen levels and multi eaves, single layered pagoda body, two layered sumeru podium.	
	铁塔 Tie Pagoda	辽宁海城 Haicheng, Liaoning	约辽末金初 About the Late Liao to Early Jin Dynasty	原高 23.6 米，现残高约 20 米 The original height is 23.6m, now about 20m.	六角七级密檐式，双层塔身，须弥座损坏不存 Hexagonal structure with seven levels and multi eaves, two layered pagoda body, sumeru podium was damaged.	1954 年修复塔基，1975 年海城地震后，对全塔修缮加固 The foundation of pagoda was repaired in 1954, and the entire pagoda was reinforced after the 1975 Haicheng Earthquake.

图版索引

49
西面塔身左侧菩萨局部
锦州广济寺塔

50
西面塔身右侧菩萨
锦州广济寺塔

51
西面塔身右侧菩萨头部
锦州广济寺塔

52
西面塔身左侧菩萨
辽宁朝阳凤凰山云接寺塔

53
南面塔身左侧菩萨局部
朝阳凤凰山云接寺塔

54
南面塔身左侧菩萨
辽宁朝阳北塔

55
南面塔身右侧菩萨
朝阳北塔

56
二层东北面塔身菩萨（正面）
辽宁喀左精严禅寺塔

57
二层东北面塔身菩萨（侧面）
喀左精严禅寺塔

58
一层东南面塔身菩萨
喀左精严禅寺塔

59
一层东南面塔身左侧菩萨局部
喀左精严禅寺塔

60
二层东南面塔身菩萨
喀左精严禅寺塔

61
二层东南面塔身左侧菩萨
喀左精严禅寺塔

62
二层东南面塔身右侧菩萨
喀左精严禅寺塔

63
二层西南面塔身菩萨（侧面）
喀左精严禅寺塔

64
一层西北面塔身菩萨
喀左精严禅寺塔

65
一层西北面塔身左侧菩萨
喀左精严禅寺塔

66
一层西北面塔身右侧菩萨
喀左精严禅寺塔

67
二层西北面塔身菩萨
喀左精严禅寺塔

68
一层西南面塔身菩萨（正面）
喀左精严禅寺塔

69
一层西南面塔身左侧菩萨
喀左精严禅寺塔

70
一层西南面塔身右侧菩萨
喀左精严禅寺塔

71
一层西南面塔身菩萨（侧面）
喀左精严禅寺塔

72
东面塔身左侧菩萨
内蒙古宁城大明塔

73
东面塔身右侧菩萨
宁城大明塔

74
东面塔身右侧菩萨局部
宁城大明塔

75
南面塔身左侧菩萨
宁城大明塔

76
南面塔身右侧菩萨
宁城大明塔

77
二层西南面塔身之一
内蒙古呼和浩特华严经塔

78
二层西南面塔身之二
呼和浩特华严经塔

79
二层西南面塔身左侧菩萨头部
呼和浩特华严经塔

80
二层西南面塔身左侧菩萨（正面）
呼和浩特华严经塔

81
二层东南面塔身右侧菩萨头部
呼和浩特华严经塔

82
二层东南面塔身右侧菩萨（侧面）
呼和浩特华严经塔

83
二层东南面塔身右侧菩萨（正面）
呼和浩特华严经塔

84
一层东北面塔身左侧菩萨
呼和浩特华严经塔

85
一层东北面塔身右侧菩萨
呼和浩特华严经塔

86
一层西南面塔身左侧菩萨
呼和浩特华严经塔

87
一层东南面塔身右侧菩萨
呼和浩特华严经塔

88
一层西北面塔身左侧菩萨
呼和浩特华严经塔

89
一层西北面塔身右侧菩萨
呼和浩特华严经塔

90
北面塔身飞天
辽宁海城金塔

91
东面塔身飞天
海城金塔

92
北面塔身左侧飞天
海城金塔

93
西面塔身右侧飞天
海城金塔

94
东面塔身左侧飞天
海城金塔

95
东面塔身右侧飞天
海城金塔

96
东南面塔身飞天
辽宁辽阳白塔

97
南面塔身飞天
辽阳白塔

98
东南面塔身左侧飞天
辽阳白塔

99
东南面塔身右侧飞天
辽阳白塔

100
南面塔身左侧飞天
辽阳白塔

101
南面塔身右侧飞天
辽阳白塔

102
西北面塔身飞天
辽阳白塔

103
东面塔身飞天
辽阳白塔

104
西北面塔身左侧飞天
辽阳白塔

105
西北面塔身右侧飞天
辽阳白塔

106
东面塔身左侧飞天
辽阳白塔

107
东面塔身右侧飞天
辽阳白塔

108
西北面塔身飞天
辽宁锦州广济寺塔

109
西北面塔身左侧飞天
锦州广济寺塔

110
西北面塔身右侧飞天
锦州广济寺塔

111
西面塔身左侧飞天
锦州广济寺塔

112
西面塔身右侧飞天
锦州广济寺塔

113
南面塔身左侧飞天
辽宁朝阳凤凰山大宝塔

114
南面塔身右侧飞天
朝阳凤凰山大宝塔

115
西面塔身左侧飞天
辽宁朝阳凤凰山云接寺塔

116
西面塔身右侧飞天
朝阳凤凰山云接寺塔

117
东面塔身左侧飞天
辽宁朝阳北塔

118
东面塔身右侧飞天
朝阳北塔

119
南面塔身左侧飞天
朝阳北塔

120
南面塔身右侧飞天
朝阳北塔

121
东面塔身左侧飞天
内蒙古宁城大明塔

122
东面塔身右侧飞天
宁城大明塔

123
南面塔身左侧飞天
宁城大明塔

124
南面塔身右侧飞天
宁城大明塔

125
西面塔身左侧飞天
宁城大明塔

126
西面塔身右侧飞天
宁城大明塔

127
一层西北面塔身
内蒙古巴林右旗庆州白塔

128
一层西南面塔身
巴林右旗庆州白塔

129
一层西北面塔身飞天（迦陵频伽）
巴林右旗庆州白塔

130
一层西南面塔身飞天（迦陵频伽）
巴林右旗庆州白塔

131
塔身飞天之一
内蒙古巴林左旗南塔
巴林左旗博物馆藏

132
塔身飞天之二
巴林左旗南塔
巴林左旗博物馆藏

133
南面塔身左侧飞天
北京房山云居寺北塔

134
南面塔身右侧飞天
房山云居寺北塔

135
二层塔身
辽宁喀左精严禅寺塔

136
二层南面塔身天王
喀左精严禅寺塔

137
二层南面塔身左侧天王
喀左精严禅寺塔

138
二层南面塔身右侧天王
喀左精严禅寺塔

139
一层西面塔身天王
喀左精严禅寺塔

140
一层西面塔身左侧天王头部
喀左精严禅寺塔

141
一层西面塔身左侧天王
喀左精严禅寺塔

142
一层西面塔身右侧天王
喀左精严禅寺塔

143
一层西面塔身右侧天王头部
喀左精严禅寺塔

144
二层西面塔身天王
喀左精严禅寺塔

145
二层西面塔身左侧天王
喀左精严禅寺塔

146
二层西面塔身右侧天王
喀左精严禅寺塔

147
二层西面塔身右侧天王头部
喀左精严禅寺塔

148
二层西面塔身右侧天王（侧面）
喀左精严禅寺塔

149
二层东面塔身左侧天王
喀左精严禅寺塔

150
二层东面塔身右侧天王
喀左精严禅寺塔

151
二层北面塔身
喀左精严禅寺塔

152
南面、东南面、东面塔身（左→右）
内蒙古宁城大明塔

153
西北面塔身
宁城大明塔

154
东南面塔身左侧天王
宁城大明塔

155
东南面塔身右侧天王
宁城大明塔

156
西南面塔身左侧天王
宁城大明塔

157
西南面塔身右侧天王
宁城大明塔

158
西北面塔身左侧天王
宁城大明塔

159
西北面塔身左侧天王头部
宁城大明塔

160
西北面塔身右侧天王头部
宁城大明塔

161
西北面塔身右侧天王
宁城大明塔

162
东北面塔身左侧天王
宁城大明塔

163
东北面塔身右侧天王
宁城大明塔

164
一层西面塔身左侧天王
内蒙古巴林右旗庆州白塔

165
一层西面塔身右侧天王
巴林右旗庆州白塔

166
二层南面塔身
内蒙古呼和浩特华严经塔

167
二层南面塔身左侧天王
呼和浩特华严经塔

168
二层南面塔身右侧天王
呼和浩特华严经塔

169
二层南面塔身右侧天王局部
呼和浩特华严经塔

170
二层西面塔身右侧天王
呼和浩特华严经塔

171
二层西面塔身左侧天王
呼和浩特华严经塔

172
二层北面塔身左侧天王
呼和浩特华严经塔

173
二层北面塔身右侧天王
呼和浩特华严经塔

174
一层北面塔身
呼和浩特华严经塔

175
一层西面塔身
呼和浩特华严经塔

176
一层北面塔身左侧天王
呼和浩特华严经塔

177
一层北面塔身右侧天王
呼和浩特华严经塔

178
一层西面塔身左侧天王
呼和浩特华严经塔

179
一层西面塔身右侧天王
呼和浩特华严经塔

180
一层东南面塔身左侧天王
呼和浩特华严经塔

181
一层南面塔身右侧天王
呼和浩特华严经塔

182
塔身和二层须弥座
辽宁海城金塔

183
二层须弥座
海城金塔

184
一层须弥座东北面北角负塔力士
海城金塔

185
一层须弥座东南面南角负塔力士
海城金塔

186
一层须弥座东北面东角负塔力士头部
海城金塔

187
一层须弥座东北面东角负塔力士
海城金塔

188
二层须弥座东北面北角负塔力士
海城金塔

189
东北角负塔力士
辽宁朝阳凤凰山云接寺塔

190
西南角负塔力士（西侧）
朝阳凤凰山云接寺塔

191
西南角负塔力士（正面）
朝阳凤凰山云接寺塔

192
西北角负塔力士
朝阳凤凰山云接寺塔

193
西南角负塔力士（南侧）
朝阳凤凰山云接寺塔

194
须弥座西面假门
辽宁朝阳北塔

195
须弥座西面假门右侧天王
朝阳北塔

196
须弥座东面假门左侧天王
朝阳北塔

197
须弥座东面假门
朝阳北塔

198
须弥座西面西北角负塔力士
辽宁喀左精严禅寺塔

199
须弥座南面东南角负塔力士
北京房山云居寺北塔

200
须弥座北面东北角负塔力士
房山云居寺北塔

201
须弥座北面西北角负塔力士
房山云居寺北塔

202
须弥座东南面乐人之一
辽宁海城金塔

203
须弥座东南面乐人之二
海城金塔

204
须弥座西面乐人之一
辽宁朝阳北塔

205
须弥座西面乐人之二
朝阳北塔

206
须弥座北面舞人之一
北京房山云居寺北塔

207
须弥座北面舞人之二
房山云居寺北塔

208
须弥座东北面舞乐人
房山云居寺北塔

209
须弥座东南面乐人
房山云居寺北塔

210
须弥座东北面乐人
房山云居寺北塔

211
须弥座西面乐人之一
房山云居寺北塔

212
须弥座西面乐人之二
房山云居寺北塔

213
须弥座西面舞人
房山云居寺北塔

214
须弥座西南面乐人
房山云居寺北塔

215
须弥座北面舞乐人
房山云居寺北塔

216
须弥座北面乐人
房山云居寺北塔

217
须弥座北面舞人之三
房山云居寺北塔

218
须弥座南面供养人
辽宁朝阳凤凰山大宝塔

219
须弥座南面供养人
北京房山云居寺北塔

220
须弥座西南面迦陵频伽之一
房山云居寺北塔

221
须弥座西南面迦陵频伽之二
房山云居寺北塔

222
须弥座北面供养人之一
辽宁海城金塔

223
须弥座北面供养人之二
海城金塔

224
须弥座东北面供养人之一
海城金塔

225
须弥座东北面供养人之二
海城金塔

226
须弥座东北面供养人之三
海城金塔

227
须弥座西南面供养人
海城金塔

PLATE INDEX

150
The King of heaven (right) on the east side of pagoda of the second floor
Pagoda of Jingyan Temple in Kazuo

151
Pagoda body of the north side of the second floor
Pagoda of jingyan Temple in Kazuo

152
Pagoda body on the south, southeast, east side (from left to right)
Daming Pagoda in Ningcheng, Inner Mongolia

153
Pagoda body on the northwest side
Daming Pagoda in Ningcheng

154
The King of heaven on the southeast side (left)
Daming Pagoda in Ningcheng

155
The King of heaven on the southeast side (right)
Daming Pagoda in Ningcheng

156
The King of heaven on the southwest side (left)
Daming Pagoda in Ningcheng

157
The King of heaven on the southwest side (right)
Daming Pagoda in Ningcheng

158
The King of heaven on the northwest side (left)
Daming Pagoda in Ningcheng

159
The head of the King of heaven on the northwest side (left)
Daming Pagoda in Ningcheng

160
The head of the King of heaven on the northwest side (right)
Daming Pagoda in Ningcheng

161
The King of heaven on the northwest side (right)
Daming Pagoda in Ningcheng

162
The King of heaven on the northeast side (left)
Daming Pagoda in Ningcheng

163
The King of heaven on the northeast side (right)
Daming Pagoda in Ningcheng

164
The King of heaven (left) on the west side of pagoda of the first floor
Qingzhou White Pagoda in Balin Youqi, Inner Mongolia

165
The King of heaven (right) on the west side of pagoda of the first floor
Qingzhou White Pagoda in Balin Youqi

166
Pagoda body of the south side of the second floor
Huayanjing Pagoda in Hohhot, Inner Mongolia

167
The King of heaven (left) on the south side of pagoda of the second floor
Huayanjing Pagoda in Hohhot

168
The King of heaven (right) on the south side of pagoda of the second floor
Huayanjing Pagoda in Hohhot

169
Part of the King of heaven (right) on the south side of pagoda of the second floor
Huayanjing Pagoda in Hohhot

170
The King of heaven (right) on the west side of pagoda of the second floor
Huayanjing Pagoda in Hohhot

171
The King of heaven (left) on the west side of pagoda of the second floor
Huayanjing Pagoda in Hohhot

172
The King of heaven (left) on the north side of pagoda of the second floor
Huayanjing Pagoda in Hohhot

173
The King of heaven (right) on the north side of pagoda of the second floor
Huayanjing Pagoda in Hohhot

174
Pagoda body on the north side of the first floor
Huayanjing Pagoda in Hohhot

175
Pagoda body on the west side of the first floor
Huayanjing Pagoda in Hohhot

176
The King of heaven (left) on the north side of pagoda of the first floor
Huayanjing Pagoda in Hohhot

177
The King of heaven (right) on the north side of pagoda of the first floor
Huayanjing Pagoda in Hohhot

178
The King of heaven (left) on the west side of pagoda of the first floor
Huayanjing Pagoda in Hohhot

179
The King of heaven (right) on the west side of pagoda of the first floor
Huayanjing Pagoda in Hohhot

180
The King of heaven (left) on the southeast side of pagoda of the first floor
Huayanjing Pagoda in Hohhot

181
The King of heaven (right) on the south side of pagoda of the first floor
Huayanjing Pagoda in Hohhot

182
Pagoda body and two-layered sumeru podium
Jin Pagoda in Haicheng, Liaoning

183
Two-layered sumeru podium
Jin Pagoda in Haicheng

184
Strong man (north corner) on the northeast side of sumeru podium of the first floor
Jin Pagoda in Haicheng

185
Strong man(south corner) on the southeast side of sumeru podium of the first floor
Jin Pagoda in Haicheng

186
The head of strong man (east corner) on the northeast side of sumeru podium of the first floor
Jin Pagoda in Haicheng

187
Strong man (east corner) on the northeast side of sumeru podium of the first floor
Jin Pagoda in Haicheng

188
Strong man (north corner) on the northeast side of sumeru podium of the second floor
Jin Pagoda in Haicheng

189
Strong man at the northeast corner
Pagoda of Yunjie Temple at Fenghuang Mountain in Chaoyang, Liaoning

190
Strong man at the southwest corner (west)
Pagoda of Yunjie Temple at Fenghuang Mountain in Chaoyang

191
Strong man at the southwest corner (front)
Pagoda of Yunjie Temple at Fenghuang Mountain in Chaoyang

192
Strong man at the northwest corner
Pagoda of Yunjie Temple at Fenghuang Mountain in Chaoyang

193
Strong man at the southwest corner (south)
Pagoda of Yunjie Temple at Fenghuang Mountain in Chaoyang

194
Blank door on the west side of sumeru podium
North Pagoda in Chaoyang, Liaoning

195
The King of heaven on the right side of blank door of sumeru podium (west)
North Pagoda in Chaoyang

196
The King of heaven on the left side of blank door of sumeru podium (east)
North Pagoda in Chaoyang

197
Blank door on the east side of sumeru podium
North Pagoda in Chaoyang

198
Strong man at the northwest corner of sumeru podium (west)
Pagoda of Jingyan Temple in Kazuo, Liaoning

199
Strong man at the southeast corner of sumeru podium (south)
North Pagoda of Yunju Temple in Fangshan, Beijing

200
Strong man at the northeast corner of sumeru podium (north)
North Pagoda of Yunju Temple in Fangshan

201
Strong man at the northwest corner of sumeru podium (north)
North Pagoda of Yunju Temple in Fangshan

202
Musician on the southeast side of sumeru podium (first)
Jin Pagoda in Haicheng, Liaoning

203
Musician on the southeast side of sumeru podium (second)
Jin Pagoda in Haicheng

204
Musicians on the west side of sumeru podium (first)
North Pagoda in Chaoyang, Liaoning

205
Musicians on the west side of sumeru podium (second)
North Pagoda in Chaoyang

206
Dancer on the north side of sumeru podium (first)
North Pagoda of Yunju Temple in Fangshan, Beijing

207
Dancer on the north side of sumeru podium (second)
North Pagoda of Yunju Temple in Fangshan

208
Dancer and musician on the northeast side of sumeru podium
North Pagoda of Yunju Temple in Fangshan

209
Musician on the southeast side of sumeru podium
North Pagoda of Yunju Temple in Fangshan

210
Musician on the northeast side of sumeru podium
North Pagoda of Yunju Temple in Fangshan

211
Musician on the west side of sumeru podium (first)
North Pagoda of Yunju Temple in Fangshan

212
Musician on the west side of sumeru podium (second)
North Pagoda of Yunju Temple in Fangshan

213
Dancer on the west side of sumeru podium
North Pagoda of Yunju Temple in Fangshan

214
Musician on the southwest side of sumeru podium
North Pagoda of Yunju Temple in Fangshan

215
Dancers and Musicians on the north side of sumeru podium
North Pagoda of Yunju Temple in Fangshan

216
Musician on the north side of sumeru podium
North Pagoda of Yunju Temple in Fangshan

217
Dancer on the north side of sumeru podium (third)
North Pagoda of Yunju Temple in Fangshan

218
Buddhist donators on the south side of sumeru podium
Dabao Pagoda at Fenghuang Mountain in Chaoyang, Liaoning

219
Buddhist donator on the south side of sumeru podium
North Pagoda of Yunju Temple in Fangshan, Beijing

220
Kalavinka on the southwest side of sumeru podium (first)
North Pagoda of Yunju Temple in Fangshan

221
Kalavinka on the southwest side of sumeru podium (second)
North Pagoda of Yunju Temple in Fangshan

222
Buddhist donator on the north side of sumeru podium (first)
Jin Pagoda in Haicheng, Liaoning

223
Buddhist donators on the north side of sumeru podium (second)
Jin Pagoda in Haicheng

224
Buddhist donator on the northeast side of sumeru podium (first)
Jin Pagoda in Haicheng

225
Buddhist donator on the northeast side of sumeru podium (second)
Jin Pagoda in Haicheng

226
Buddhist donator on the northeast side of sumeru podium (third)
Jin Pagoda in Haicheng

227
Buddhist donator on the southwest side of sumeru podium
Jin Pagoda in Haicheng

228
Buddhist donator on the southwest side of sumeru podium (first)
North Pagoda of Yunju Temple in Fangshan, Beijing

229
Buddhist donator on the southwest side of sumeru podium (second)
North Pagoda of Yunju Temple in Fangshan

230
Buddhist donators and musician on the sumeru podium
North Pagoda of Yunju Temple in Fangshan

231
Animal rider on the southeast side of sumeru podium
North Pagoda of Yunju Temple in Fangshan

232
Buddhist donators on the south side of sumeru podium (first)
North Pagoda of Yunju Temple in Fangshan

233
Buddhist donators on the south side of sumeru podium (second)
North Pagoda of Yunju Temple in Fangshan

234
Picture of praying to Buddha on the west side of sumeru podium
Pagoda of Jingyan Temple in Kazuo, Liaoning

235
Picture of praying to Buddha on the southeast side of sumeru podium
Pagoda of Jingyan Temple in Kazuo

236
Guardian lions on the west side (left) and southwest side (right) of sumeru podium
Pagoda of Jingyan Temple in Kazuo

237
Guardian lion on the southwest side of sumeru podium
Pagoda of Jingyan Temple in Kazuo

238
Guardian lion on the west side of sumeru podium
Pagoda of Jingyan Temple in Kazuo

239
Guardian lion on the northwest side of sumeru podium (first)
North Pagoda of Yunju Temple in Fangshan, Beijing

240
Guardian lion on the southwest side of sumeru podium
North Pagoda of Yunju Temple in Fangshan

241
Dancer and lion on the west side of sumeru podium (first)
North Pagoda of Yunju Temple in Fangshan

242
Dancer and lion on the west side of sumeru podium (second)
North Pagoda of Yunju Temple in Fangshan

243
Guardian lion on the northeast side of sumeru podium
North Pagoda of Yunju Temple in Fangshan

244
Guardian lion on the northwest side of sumeru podium (second)
North Pagoda of Yunju Temple in Fangshan

245
Guardian lion on the northeast side of sumeru podium (first)
Jin Pagoda in Haicheng, Liaoning

246
Guardian lion on the northeast side of sumeru podium (second)
Jin Pagoda in Haicheng

247
Guardian lion on the east side of sumeru
podium
Jin Pagoda in Haicheng

248
Guardian lion on the south side of sumeru
podium
Jin Pagoda in Haicheng

249
Guardian lion on the north side of sumeru
podium
Jin Pagoda in Haicheng

250
Guardian lion on the southeast side of sumeru
podium
Jin Pagoda in Haicheng

251
The head of guardian lion on the southeast
side of sumeru podium
Jin Pagoda in Haicheng

彩页（左—右）

Color Pages (Left–Right)

1. 巴林左旗南塔 South Pagoda in Balin Zuoqi
2. 大明塔 Dcming Pagoda
3. 华严经塔 Huayanjing Pagoda
4. 庆州白塔 Qingzhou White Pagoda

5. 精严禅寺塔 Pagoda of Jingyan Temple
6. 朝阳北塔 North Pagoda in Chaoyang
7. 朝阳南塔 South Pagoda in Chaoyang
8. 大宝塔 Dabao Pagoda
9. 云接寺塔 Pagoda of Yunjie Temple
10. 八棱观塔 Balengguan Pagoda

11. 广济寺塔 Pagoda of Guangji Temple
12. 辽阳白塔 White Pagoda in Liaoyang
13. 海城金塔 Jin Pagoda in Haicheng
14. 海城铁塔 Tie Pagoda in Haicheng
15. 铁岭白塔 White Pagoda in Tieling
16. 天成寺塔 Pagoda of Tiancheng Temple
17. 蓟县白塔 White Pagoda in Jixian
18. 云居寺北塔 North Pagoda of Yunju Temple

封面：广济寺塔佛头像

Cover: The head of Buddha in the Pagoda of Guangji Temple

封底：精严禅寺塔菩萨像

Back Cover: The Bodhisattva in the Pagoda of Jingyan Temple

彩页

Color Page

1．金塔

　　Jin Pagoda

2．金塔局部之一

　　The part of Jin Pagoda

3．金塔局部之二

　　The part of Jin Pagoda

4．辽上京南塔

　　The South Pagoda in Liao Upper Capital

5．精严禅寺塔

　　The Pagoda of Jingyan Temple